Proceedings in Life Sciences

Functions of Glutathione in Liver and Kidney

Edited by
H. Sies and A. Wendel

With 94 Figures

Springer-Verlag
Berlin Heidelberg New York 1978

Professor Dr. HELMUT SIES
Institut für Physiologische Chemie, Physikalische Biochemie
und Zellbiologie der Universität München
Goethestraße 33, 8000 München, FRG

Dozent Dr. ALBRECHT WENDEL
Physiologisch-chemisches Institut der Universität Tübingen
Hoppe-Seyler-Straße 1, 7400 Tübingen, FRG

The abstracts of the papers of this Conference have been published in "Hoppe-Seyler's Z. Physiol. Chem. *349*, 1031–1040, 1978".

ISBN-13: 978-3-642-67134-0 e-ISBN-13: 978-3-642-67132-6
DOI: 10.1007/978-3-642-67132-6

Library of Congress Cataloging in Publication Data. Gesellschaft für Biologische Chemie. Functions of glutathione in liver and kidney. (Proceedings in life sciences.) Papers from 25. Konferenz der Gesellschaft für Biologische Chemie, held at Schloß Reisensburg, Ger., July 8–11, 1978. Includes bibliographies and index. 1. Glutathione metabolism—Congresses. 2. Liver—Congresses. 3. Kidneys—Congresses. I. Sies, Helmut. II. Wendel, Albrecht, 1943—. III. Title. QP552.G58G47 1978 599'.01'2 78-21535.

Softcover reprint of the hardcover 1st edition 1978

Preface

The serene phrase, *Lest I forget thee, glutathione...*, coined by the Kosowers (1) to describe the state in the 1960's, must be replaced now by something like "Inevitable GSH" in order to characterize the current situation. The surge in interest on the ubiquitous tripeptide has been amazing, with publications on GSH running at rates as high as one per day, so that it seemed appropriate to convene international experts for a discussion of recent developments this year. Unlike the two previous meetings in this decade held in Tübingen in 1973 (2) and in Santa Ynez in 1975 (3), the scope was restricted to *Functions of Glutathione in Liver and Kidney*. Only in this way did an in-depth discussion of the current state of knowledge in a limited topic appear possible.

The last couple of years have seen a fascinating productivity in the fields of (a) *Regulation of the Glutathione Level in the Liver*, (b) *Role of γ-Glutamyltransferase in Glutathione Turnover* with emphasis on the renal enzyme, and a critical appraisal of the *γ-Glutamyl Cycle*, (c) *Hydroperoxide and Disulfide Metabolism*, enriched by the discovery of the nonselenium-dependent glutathione peroxidase activity and its relation to the glutathione-S-transferases, and the participation of the 2GSH/GSSG system in redox transitions in intact organ, cells and isolated mitochondria, and (d) a multitude of *Pharmacological and Toxicological Aspects* related to glutathione, mainly centered on the events leading to liver damage and the protective mechanisms. These topics were the subject matter for the meeting held as *25. Konferenz der Gesellschaft für Biologische Chemie* at Schloß Reisensburg, Germany, July 8-11, 1978.

For the sake of rapid publication of the papers containing, in most instances, brand-new information, we refrain from including the very extensive and lively discussions. Dedication and interaction of the renowned as well as the younger researchers surpassed all our expectations, and we hope that the book will transmit some of this fascination.

Autumn, 1978

HELMUT SIES
ALBRECHT WENDEL

---

1. Kosower, E.M., Kosower, N.S.: Nature (London) 224, 117-120 (1969)
2. Flohé, L., Benöhr, C., Sies, H., Waller, D., Wendel, A. (eds.): Glutathione. Stuttgart: Thieme 1974
3. Arias, I.M., Jakoby, W.B. (eds.): Glutathione: Metabolism and Function. New York: Raven Press 1976

Contents

Chapter I
The Regulation of the Glutathione Level in the Liver 1

Turnover of Glutathione in Rat Liver
N. TATEISHI and T. HIGASHI (With 3 Figures) 3

Regulation of the Hepatic Concentration of Reduced Glutathione
H.A. KREBS, R. HEMS, and J. VIÑA 8

The Role of the Cystathionine Pathway in Glutathione Regulation by Isolated Hepatocytes
D.J. REED and P.W. BEATTY (With 2 Figures) 13

Physiological Significance of Elevation of Liver γ-Glutamyltranspeptidase in Late Pregnancy
T. HIGASHI and N. TATEISHI (With 3 Figures) 22

Glutathione Efflux from Perfused Rat Liver and Its Relation to Glutathione Uptake by the Kidney
G.M. BARTOLI, D. HÄBERLE, and H. SIES 27

Unidirectional Transport of Reduced Glutathione in Rat Liver and Its Metabolization in the Extracellular Space
R. HAHN and W. OBERRAUCH (With 5 Figures) 32

Chapter II
The Role of γ-Glutamyl-Transferase in Glutathione Turnover 41

Current Status of the γ-Glutamyl Cycle
A. MEISTER (With 3 Figures) 43

Topology and Function of Renal γ-Glutamyl Transpeptidase
S. SILBERNAGL, W. PFALLER, H. HEINLE, and A. WENDEL (With 5 Figures) 60

The Membrane Association and Physiological Function of Rat Renal γ-Glutamyltranspeptidase
N.P. CURTHOYS, R.P. HUGHEY, and P.J. COYLE (With 5 Figures) 70

Studies on the Active Site of γ-Glutamyltransferase
J.S. ELCE 78

The Role of Glutathione in Transport Activity in Kidney
F.H. LEIBACH, D.J. PILLION, J. MENDICINO, and D. PASHLEY 84

Chapter III
Hydroperoxide and Disulfide Metabolism 93

Hydroperoxide Metabolism: An Overview
B. CHANCE, A. BOVERIS, Y. NAKASE, and H. SIES (With 9 Figures) 95

The Selenium Moiety of Glutathione Peroxidase
A. WENDEL, B. KERNER, and K. GRAUPE (With 3 Figures) 107

Non-Selenium-Dependent Glutathione Peroxidase
R.F. BURK and R.A. LAWRENCE (With 4 Figures) 114

Properties of Glutathione Disulfide (GSSG) and Glutathione-S-Conjugate Release from Perfused Rat Liver
H. SIES, A. WAHLLÄNDER, and Ch. WAYDHAS (With 5 Figures) 120

Reduction of Diazenes and Hydroperoxides by Rat Liver Mitochondria
P.C. JOCELYN (With 11 Figures) 127

Effect of Oxidation of Glutathione and Membrane Thiol Groups on Mitochondrial Functions
N. SILIPRANDI, D. SILIPRANDI, A. BINDOLI, and A. TONINELLO (With 5 Figures) 139

The Roles of Glutathione, Thioltransferase, and Glutathione Reductase in the Scission of Sulfur-Sulfur Bonds
B. MANNERVIK and K. AXELSSON 148

Chapter IV
Pharmacological and Toxicological Aspects 155

The Glutathione Transferases in Detoxification
W.B. JAKOBY (With 1 Figure) 157

Functions of Glutathione in Drug Metabolism
S. ORRENIUS and D.P. JONES (With 8 Figures) 164

Lipid Peroxidation Induced by Ethanol and Halogenated Hydrocarbons in vivo as Measured by Ethane Exhalation
H. KAPPUS, U. KÖSTER, D. KÖSTER-ALBRECHT, H. KIECZKA, and H. REMMER (With 3 Figures) 176

Drug-induced Lipid Peroxidation in Mouse Liver
A. WENDEL, S. FEUERSTEIN, and K.-H. KONZ (With 3 Figures) 183

Cellular Functions in Glutathione-Depleted Hepatocytes
J. HÖGBERG, I. ANUNDI, A. KRISTOFERSON, and A.H. STEAD (With 4 Figures) 189

Glutathione and Glutathione Conjugate Metabolism in Isolated Liver and Kidney Cells
D.P. JONES, A.H. STEAD, P. MOLDÉUS, and S. ORRENIUS (With 9 Figures) 194

Glutathione-Requiring Enzymes in the Metabolism of Prostaglandin Endoperoxides
E. CHRIST-HAZELHOF and D.H. NUGTEREN (With 3 Figures) 201

Subject Index 207

GSH

List of Contributors

ANUNDI, I., Department of Forensic Medicine, Karolinska Institutet, S-10401 Stockholm 60, Sweden

AXELSSON, K., Department of Biochemistry, Arrhenius Laboratory, University of Stockholm, S-10691 Stockholm, Sweden

BARTOLI, G.M., Istituto di Patologia Generale, Universita Cattolica, Via della Pineta Sacchetti, 644, 00168-Roma, Italy

BEATTY, P.W., Department of Biochemistry and Biophysics, Oregon State University, Corvallis, OR 97331, USA

BINDOLI, A., Centro per lo Studio della Fisiologia Mitocondriale, Via Marzolo 3, 35100 Padova, Italy

BOVERIS, A., Johnson Research Foundation, University of Pennsylvania, Philadelphia, PA 19104, USA

BURK, R.F., Department of Medicine, LSU Medical Center, Shreveport, LA 71130, USA

CHANCE, B., Johnson Research Foundation, University of Pennsylvania, Philadelphia, PA 19104, USA

CHRIST-HAZELHOF, E., Unilever Research Vlaardingen, P.O. Box 114, 3130 AC Vlaardingen, The Netherlands

COYLE, P.J., Department of Biochemistry, University of Pittsburgh School of Medicine, Pittsburgh, PA 15261, USA

CURTHOYS, N.P., Department of Biochemistry, University of Pittsburgh, School of Medicine, Pittsburgh, PA 15261, USA

ELCE, J.S., Department of Biochemistry, Queen's University, Kingston, Ontario, K7L 3N6, Canada

FEUERSTEIN, S., Physiologisch-Chemisches Institut der Universität Tübingen, Hoppe-Seyler-Str. 1, 7400 Tübingen, FRG

GRAUPE, K., Physiologisch-Chemisches Institut der Universität Tübingen, Hoppe-Seyler-Str. 1, 7400 Tübingen, FRG

HAHN, R., Physiologisch-Chemisches Institut der Universität Tübingen, Hoppe-Seyler-Str. 1, 7400 Tübingen, FRG

HÄBERLE, D., Institut für Physiologie, Universität München, Pettenkoferstr. 12, 8000 München 2, FRG

HEINLE, H., Physiologisch-Chemisches Institut der Universität Tübingen, Hoppe-Seyler-Str. 1, 7400 Tübingen, FRG

HEMS, R., Metabolic Research Laboratory, Nuffield Department of Clinical Medicine, Radcliffe Infirmary, Oxford, England

HIGASHI, T., Department of Biochemistry, Institute for Cancer Research, Osaka University Medical School, Osaka, Japan

HÖGBERG, J., Department of Forensic Medicine, Karolinska Institutet, 10401 Stockhom 60, Sweden

HUGHEY, R.P., Department of Biochemistry, University of Pittsburgh, School of Medicine, Pittsburgh, PA 15261, USA

JAKOBY, W.B., National Institutes of Health, Building 10/Room 9N-119, Bethesda, MD 20014, USA

JOCELYN, P.C., Department of Biochemistry, University of Edinburgh Medical School, Teviot Place, Edinburgh EH8 9AG, Scotland

JONES, D.P., Department of Forensic Medicine, Karolinska Institutet, 10401 Stockholm 60, Sweden

KAPPUS, H., Medizinisches Institut für Lufthygiene und Silikoseforschung an der Universität Düsseldorf, Gurlittstraße 53, 4000 Düsseldorf, FRG

KERNER, B., Physiologisch-Chemisches Institut der Universität Tübingen, Hoppe-Seyler-Str. 1, 7400 Tübingen, FRG

KIECZKA, H., Institut für Toxikologie, Universität Tübingen, Wilhelmstraße 56, 7400 Tübingen, FRG

KÖSTER, U., Institut für Toxikologie, Universität Tübingen, Wilhelmstraße 56, 7400 Tübingen, FRG

KÖSTER-ALBRECHT, D., Institut für Toxikologie, Universität Tübingen, Wilhelmstraße 56, 7400 Tübingen, FRG

KREBS, Sir H.A., Metabolic Research Laboratory, Nuffield Department of Clinical Medicine, Radcliffe Infirmary, Oxford, England

KRISTOFERSON, A., Department of Forensic Medicine, Karolinska Institutet, 10401 Stockholm 60, Sweden

KONZ, K.-H., Physiologisch-Chemisches Institut, Hoppe-Seyler-Str. 1, 7400 Tübingen, FRG

LAWRENCE, R.A., Department of Medicine, LSU Medical Center, Shreveport, LA 71130, USA

LEIBACH, F.H., Department of Cell and Molecular Biology, Medical College of Georgia, Augusta, GA 30901, USA

MANNERVIK, B., Department of Biochemistry, Arrhenius Laboratory, University of Stockholm, 10691 Stockholm, Sweden

MEISTER, A., Department of Biochemistry, Cornell University Medical College, 1300 York Avenue, New York, NY 10021, USA

MENDICINO, J., Department of Cell and Molecular Biology, Medical College of Georgia, Augusta, GA 30901, USA

MOLDÉUS, P., Department of Forensic Medicine, Karolinska Institutet, 10401 Stockholm 60, Sweden

NAKASE, Y., Johnson Research Foundation, University of Pennsylvania, Philadelphia, PA 19104, USA

NUGTEREN, D.H., Unilever Research Vlaardingen, P.O. Box 114, 3130 AC Vlaardingen, The Netherlands

OBERRAUCH, W., Physiologisch-Chemisches Institut der Universität Tübingen, Hoppe-Seyler-Str. 1, 7400 Tübingen, FRG

ORRENIUS, S., Department of Forensic Medicine, Karolinska Institutet, 10401 Stockholm 60, Sweden

PASHLEY, D., Department of Cell and Molecular Biology, Medical College of Georgia, Augusta, GA 30901, USA

PFALLER, W., Physiologisches Institut der Universität Innsbruck, Fritz-Pregl-Str. 3, 6010 Innsbruck, Austria

PILLION, D.J., Department of Cell and Molecular Biology, Medical College of Georgia, Augusta, GA 30901, USA

REED, D.J., Department of Biochemistry and Biophysics, Oregon State University, Corvallis, OR 97331, USA

REMMER, H., Institut für Toxikologie, Universität Tübingen, Wilhelmstraße 56, 7400 Tübingen, FRG

SIES, H., Institut für Physiologische Chemie, Physikalische Biochemie und Zellbiologie, Goethestr. 33, 8000 München 2, FRG

SILBERNAGL, S., Physiologisches Institut der Universität Innsbruck, Fritz-Pregl-Str. 3, 6010 Innsbruck, Austria

SILIPRANDI, D., Centro per lo Studio della Fisiologia Mitocondriale, Via Marzolo 3, 35100 Padova, Italy

SILIPRANDI, N., Centro per lo Studio della Fisiologia Mitocondriale, Via Marzolo 3, 35100 Padova, Italy

STEAD, A.H., Department of Forensic Medicine, Karolinska Institutet, 10401 Stockholm 60, Sweden

TATEISHI, N., Department of Biochemistry, Institute for Cancer Research, Osaka University Medical School, Osaka, Japan

TONINELLO, A., Centro per lo Studio della Fisiologia Mitocondriale, Via Marzolo 3, 35100 Padova, Italy

VIÑA, J., Metabolic Research Laboratory, Nuffield Department of Clinical Medicine, Radcliffe Infirmary, Oxford, England

WAHLLÄNDER, A., Institut für Physiologische Chemie, Physikalische Biochemie und Zellbiologie, Goethestr. 33, 8000 München 2, FRG

WAYDHAS, C., Institut für Physiologische Chemie, Physikalische Biochemie und Zellbiologie, Goethestr. 33, 8000 München 2, FRG

WENDEL, A., Physiologisch-Chemisches Institut der Universität Tübingen, Hoppe-Seyler-Str. 1, 7400 Tübingen, FRG

Chapter I
The Regulation of the Glutathione Level in the Liver

Turnover of Glutathione in Rat Liver

N. TATEISHI and T. HIGASHI

Summary

The glutathione content of rat liver is usually 7 to 8 μmol/g. After starving rats for one day, glutathione decreases to between two-thirds and half the normal level. When starved rats are fed a diet containing sulfur-containing amino acids, the liver glutathione level rises within a few hours. The mechanism and physiological significance of this rapid metabolic turnover of glutathione have not been elucidated. We have clearly shown that there are two pools of glutathione in rat liver, with apparently different half-lives (1.7 h and 28.5 h). The "labile" pool of glutathione functions as a reservoir of cysteine. Glutathione releases cysteine for protein synthesis when other conditions are fulfilled and the amount of cysteine becomes rate-limiting. Conversely, when excess cysteine is supplied the animals can avoid the harmful effect of excess cysteine (cystine) by storing it as liver glutathione.

A. Dependency of the Hepatic Glutathione Level on Food Intake and the Cysteine Content of the Food

When starved rats were fed a normal diet (laboratory chow), significant increases of glutathione were observed within two hours, and glutathione reached a maximum after 8 h (Fig. 1). The cysteine level in the liver also increased, but only transiently at the beginning of the feeding period (Fig. 1). The maximal level of glutathione attained depended on the amount of food ingested, with up to 10 g/day/150 g body wt. Smaller amounts of food intake always resulted in a decrease in the glutathione content, indicating a rapid turnover of glutathione (1). The quantita-

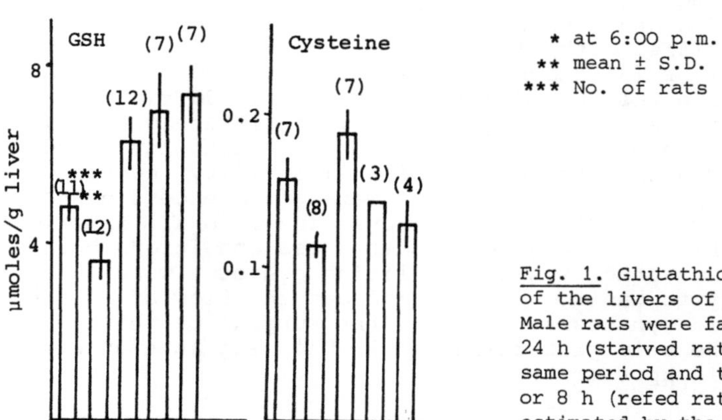

Fig. 1. Glutathione and cysteine contents of the livers of starved and refed rats. Male rats were fasted from 06:00 h. (C) for 24 h (starved rats = F) or fasted for the same period and then given food for 2, 4 or 8 h (refed rats = R). Glutathione was estimated by the method of Saville (4). Cysteine was determined by the method of Gaitonde (5)

tive dependency of the increase in glutathione on available cysteine was established using a protein-free diet fortified with a physiological amount of dietary cysteine. This relation can be explained by the enzymatic properties of glutathione-synthesizing enzymes and the cysteine concentration in the liver. The amounts of these enzymes did not change under the conditions we used. Of the constituent amino acids of glutathione, cysteine is probably the amino acid that limits the overall rate of glutathione synthesis, because it was always maintained at a low level (around 2×10^{-4}M) and the synthesizing system has a rather high K_m value (2.5×10^{-3}M) for cysteine (1). An increased supply of cysteine from the food should naturally increase glutathione synthesis.

B. Attenuating Effects of Other Amino Acids in the Diet (Especially L-Tryptophan), on the Increase of Liver Glutathione

Similar dependency of the increase of glutathione on cysteine intake was observed using a gelatin-diet supplemented with various amounts of cysteine (2). However, addition of protein to the diet resulted in a smaller increase of liver glutathione than that observed with protein-free diet, unless excess cysteine was supplied. This attenuating effect of gelatin was more clearly demonstrated when tryptophan was added to the diet. The increase of glutathione in rats given a gelatin-diet containing 0.18% cysteine was almost completely abolished by the further addition of 0.18% tryptophan, and only the addition of an unphysiologically large amount of cysteine was able to overcome the effect of tryptophan in suppressing an increase in glutathione.

C. Metabolic Fate of Dietary Cysteine

The effects of other amino acids, especially tryptophan, on the increase of glutathione previously described suggested that the presence of other amino acids affected the distribution of dietary cysteine in the body and altered the metabolic fate of cysteine.

For examination of the metabolic fate of cysteine derived from the food, rats were fasted for 40 h and then fed on diets containing gelatin as the protein source fortified with (1) cysteine (0.18%) (2) cysteine and tryptophan (0.18% each) or (3) excess cysteine (0.54%) and tryptophan (0.18%). L-[^{35}S]-cysteine was included in these diets. Addition of tryptophan to the diet suppressed the increase of liver glutathione. Tryptophan stimulated the incorporation of [^{35}S]-cysteine into liver and serum proteins, and diminished the amount of radioactivity recovered in liver glutathione and cysteine. When rats were given excess cysteine with tryptophan, the incorporation of [^{35}S]-cysteine into all fractions was larger. This probably indicates that under these conditions both glutathione synthesis and other systems utilizing cysteine were fully active. Addition of tryptophan to the diet induced the flow of dietary cysteine towards protein synthesis, and only the surplus cysteine was incorporated into the hepatic glutathione pool. This explanation was confirmed by additional experiments (2).

D. Transfer of Cysteine from Glutathione to Protein

The next problem was whether cysteine, once incorporated into glutathione, can be released for other purposes, in particular for protein synthesis.

Rats were maintained on an 18% gelatin diet for 4 days to reduce protein synthesis and their glutathione level. Then they were starved for one day, and given the first diet, that is, 42% gelatin diet fortified with ^{35}S-labeled cysteine, but still without tryptophan. A 42% gelatin diet was used here to give the rats sufficient gelatin in a short time. Four hours later, the rats were divided into two groups and they were given the second diet: one group was given a plain gelatin diet and the other group the same diet fortified with tryptophan. Then 11 h later, the distribution of ^{35}S in the liver and serum fractions was examined.

Within 4 h after the first diet, the liver glutathione level must have increased to 7 μmol/g or more judging from the experiments previously described. Eleven hours after the second diet, the group given a plain gelatin diet had 6.5 μmol of glutathione per g liver. However, the rats given tryptophan-containing diet had a significantly lower level of glutathione of 3.85 μmol/g liver. So tryptophan decreased the glutathione level.

Table 1 shows the distribution of ^{35}S in the liver and plasma of the two groups. The liver homogenate was similarly labeled with ^{35}S in both groups; but the intracellular distributions of radioactivity were en-

Table 1. Mobilization of cysteine moiety of liver glutathione by the administration of tryptophan

	Fortified amino acids	
The 1st diet (4 h)	Cysteine	Cysteine
The 2nd diet (11 h)	none	tryptophan
	μmol/g liver	
Glutathione	6.50	3.85
Cysteine	0.10	0.08
	x 10^{-3} cpm/g liver	
Liver		
Whole homogenate	358	357
Acid soluble fr.	197	129
Glutathione	156	91
Cysteine	3.2	2.0
Acid precipitable fr.	151	202
	x 10^{-3} cpm/ml of serum	
Serum		
Acid precipitable fr.	153	229
Albumin	58	101

Rats were given an 18% gelatin diet for 4 days. After starvation for 24 h, rats were given a 42% gelatin diet containing 0.09% cysteine and 10 μCi of L-[^{35}S] cysteine. Four hours later, rats were given 42% gelatin diet or the same amount of the diet fortified with 0.18% of tryptophan. The rats were killed 11 h after the second diet was given.

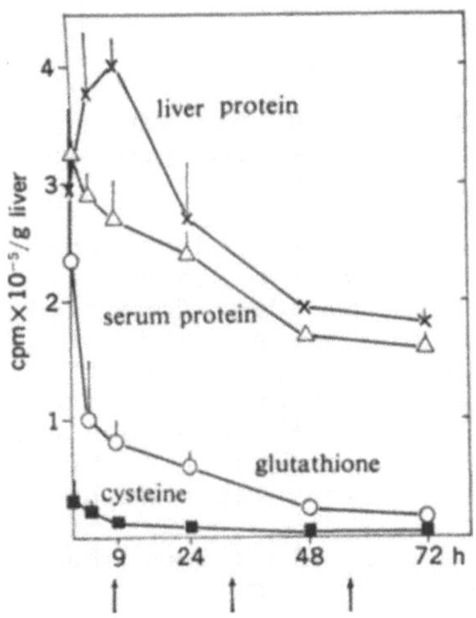

Fig. 2. Incorporation of dietary [^{35}S]-cysteine into liver and serum fractions under normal feeding conditions. Rats were starved for 9 h and then given an 18% gelatin diet containing 0.18% each of L-cysteine and L-tryptophan and 30 μCi of L-[^{35}S]-cysteine. The rats were given the same diet without radioisotopes on day 2, 3, and 4. The first group of rats was killed 15.5 h (time=0) after giving the L-[^{35}S]-cysteine-containing diet. The other group of rats was killed 3.5, 9, 24, 48, and 72 h later, and the radioactivities of their liver glutathione and of other fractions were estimated. *Arrows* indicate the time when the nonradioactive diet was given

tirely different. The tryptophan-containing diet resulted in lower labeling of glutathione and higher labeling of the acid-precipitable fraction. The pool sizes of cysteine in the liver and plasma are very small, so it seems likely that most of the increased radioactivity we found in the liver and serum proteins came from liver glutathione. This transfer of cysteine from glutathione to proteins was also suggested by another type of experiment discussed below.

Starved rats were given a diet containing [^{35}S]-cysteine and then a diet containing unlabeled cysteine for three days, and the biological decay of radioactivity was examined under normal feeding conditions. Results showed that the radioactivity in liver glutathione first decayed rapidly, and then much more slowly. Concomitant with the initial rapid decrease in the radioactivity in glutathione, labeling of liver protein with [^{35}S]-cysteine increased (Fig. 2), indicating the possible incorporation of cysteine derived from glutathione into protein.

E. Apparent Half-Lives of the Cysteine Moiety in the Two Pools of Glutathione

The semilogarithmic plot of the decay curve of glutathione can be approximated by two straight lines which may represent the turnover rates of two pools of glutathione. The apparent half-lives of these pools were estimated to be 1.7 h and 28.5 h (Fig. 3). Successive labeling of glutathione with ^{35}S-cysteine and then with ^{3}H-cysteine and examination of their differential decays confirmed the presence of two pools of glutathione (3). The low level of glutathione observed after starvation for one day did not decrease below 3 - 4 μmol/g on prolonged starvation, although the liver weight and body weight continued to decrease gradually (2). This stable pool of glutathione is probably the pool with a half-life of 28.5 h under normal feeding conditions. This pool of glutathione must serve for many reactions requiring sulfhydryl compounds in the liver. The cysteine moiety of liver glutathione with the shorter half-life is mobilized for protein synthesis when other conditions are fulfilled and the amount of cysteine becomes rate-limiting.

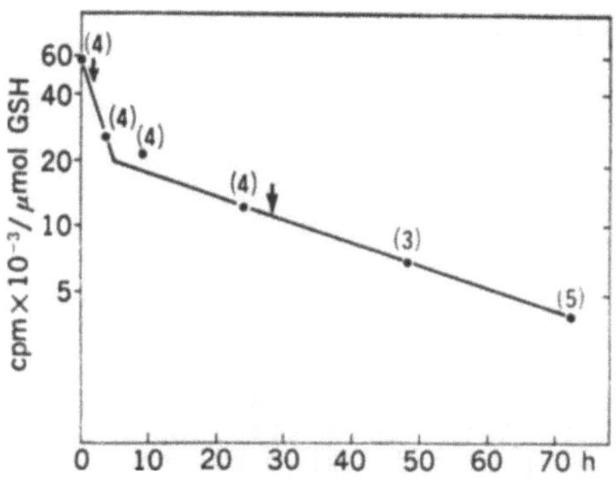

Fig. 3. Biological decay of [^{35}S]-glutathione in rat liver. The decay curve for glutathione in Figure 2 was replotted on semilogarithmic scale. *Figures in parentheses* indicate numbers of rats. *Arrows* show the half-lives of two pools of [^{35}S]-glutathione

F. How Much Liver Glutathione Can be Mobilized as a Source of Cysteine for Protein Synthesis?

We suppose that the lowest content of hepatic glutathione that rats can tolerate is about 3 μmol/g, which constitutes the "stable" pool of glutathione. The "labile" pool is readily affected by nutritional conditions. When animals normally eat laboratory chow, the labile pool is estimated to be 4 - 5 μmol/g at a maximum. On fasting, the labile glutathione will be consumed with the release of the cysteine moiety for protein synthesis.

G. Conclusions

1. There are at least two pools of glutathione in rat liver.
2. Their apparent half-lives are 1.7 h and 28.5 h.
3. The cysteine moiety of "the liver glutathione with a shorter half-life" is mobilized for protein synthesis when other conditions are fulfilled and the amount of cysteine becomes rate-limiting. We conclude that one of the functions of liver glutathione is to act as a reservoir of cysteine.

References

1. Tateishi, N., Higashi, T., Shinya, S., Naruse, A., Sakamoto, Y.: Studies on the regulation of glutathione level in rat liver. J. Biochem. 75, 93-103 (1974)
2. Tateishi, N., Higashi, T., Naruse, A., Nakashima, K., Shiozaki, H., Sakamoto, Y.: Rat liver glutathione: Possible role as a reservoir of cysteine. J. Nutr. 107, 51-60 (1977)
3. Higashi, T., Tateishi, N., Naruse, A., Sakamoto, Y.: A novel physiological role of liver glutathione as a reservoir of L-cysteine. J. Biochem. 82, 117-124 (1977)
4. Saville, B.: A scheme for the colorimetric determination of microgram amounts of thiols. Analyst 83, 670-672 (1958)
5. Gaitonde, M.K.: A spectrophotometric method for the direct determination of cysteine in the presence of other naturally occurring amino acids. Biochem. J. 104, 627-633 (1967)

Regulation of the Hepatic Concentration of Reduced Glutathione

H. A. KREBS, R. HEMS, and J. VIÑA

Summary

1. During the standard procedure for the preparation of rat hepatocytes, about half of the cellular GSH (reduced glutathione) is lost.
2. This loss is prevented by the addition of 0.1 mM EGTA (but not EDTA) to the perfusion medium.
3. On incubation with and without EGTA, isolated hepatocytes prepared in the presence of EGTA lose GSH. This loss is prevented by near-physiological concentrations of methionine or homocysteine, but not of cysteine.
4. Cysteine, at concentrations above 0.2 mM, causes a loss of GSH. Cysteine also caused a loss of adenine nucleotides and of biosynthetic capacity.
5. Serine together with methionine or homocysteine increases GSH above the value in cells from starved rats in vivo.
6. It is suggested that the deleterious effects of cysteine are caused by hydrogen peroxide formation during the intracellular oxidation of cysteine and/or the formation of mixed disulphides.
7. The beneficial effects of methionine or homocysteine in maintaining GSH may be due to a gradual release of cysteine via the cystathionine pathway, needed for the resynthesis of GSH.

It is well established that isolated hepatocytes, prepared by the principle of Berry and Friend (1969) (i.e., by treatment of the isolated perfused liver with collagenase) are a valuable material in the study of liver metabolism. Such cells, however, still have shortcomings, for instance an inability to synthesize glycogen from glucose when the external concentration of glucose is below say, 13 mM. This is in contrast with the liver in vivo which readily deposits glycogen at lower glucose concentration. Our previous experiences have shown that sub-normal performance can be due to loss of low-molecular-weight cell constituents, such as glutamate (Cornell et al., 1974) or methionine (Krebs et al., 1976). We therefore tested systematically for the loss of other low molecular weight cell constituents. This prompted us to study the intracellular concentrations of GSH in isolated hepatocytes.

As shown in Table 1, the concentration of GSH was found to fall from 5.31 mM to 3.30 mM during the preparation of the cells by our standard procedure (Cornell et al., 1974; Krebs et al., 1974). There was a further fall to 2.47 mM after 60 min incubation. The fall during the preparation of the cells was greatly decreased by the addition to the perfusion medium of 0.1 mM EGTA (but not EDTA). However cells prepared in the presence of EGTA still lost GSH when incubated in the presence of EGTA in the standard incubation medium. EGTA was first used in the preparation of hepatocytes by Seglen (1972).

Addition of lactate (10 mM), oleate (1 mM) or glutamine (5 mM) with or without EDTA (0.1 mM) had no effect on the loss of GSH. The addition of 0.1 mM EGTA, as already mentioned, prevented the fall during the preparation; but it did not do so during incubation, even after the

Table 1. [GSH] in rat liver preparations

Material	[GSH]
Freeze-clamped, fed rats	5.31
Freeze-clamped, 48 h fasted rats	2.68
Isolated hepatocytes (fed rats)	
Freshly prepared, standard method	3.30
Same cells after 60 min incubation	2.47
Freshly prepared with 0.1 mM EGTA	4.38
Same cells after 60 min incubation	2.46

Standard method refers to that described by Cornell et al. (1974) and Krebs et al. (1974). Prepared with EGTA refers to a modified procedure of Seglen (1972).

Table 2. Effect of methionine on the maintenance of [GSH] in the presence of EGTA after 60 min incubation (fed rat)

Initial [GSH] 4.38 µmol/g		
[Methionine] mM	[GSH] µmol/g	% loss
1.0	4.10	5
0.50	3.80	13
0.20	3.87	11
0.10	2.94	33
0.05	2.56	41
0	2.46	43

addition of various substrates (glucose, glutamine, ethanol, lysine, dithiothreitol).

It is known from the work of Tateishi et al. (1974) that the concentration of GSH decreases in vivo on fasting. A similar difference was found in hepatocytes isolated from fed and fasted rats. The respective values (Table 1) for hepatocytes from fed and fasted rats were 4.7 mM and 2.17 mM.

Among a large number of substances tested, only methionine and homocysteine maintained the concentration of GSH during an incubation period of 60 min. Ethionine was inactive. Methionine was effective at 0.2 mM and above (Table 2). The normal concentration of methionine in the plasma is about 0.04 mM and in the liver about 0.1 mM. The effect of methionine was increased by serine, while serine alone had no effect (Table 3). This is of relevance to the interpretation of the methionine effect. Homocysteine was effective at somewhat lower concentrations; 0.06 mM had a major effect (Table 4).

It might be expected that cysteine is a limiting factor in the synthesis of GSH and that the addition of cysteine may therefore raise the intracellular GSH concentration. However, contrary to the expectation, addition of cysteine, even at 0.25 mM (the approximate tissue concentration of cysteine) caused the destruction of the intracellular GSH (Table 5). When the added cysteine was 4 mM, GSH was no longed detectable in the cells after 60 min incubation.

Table 3. Effect of methionine and serine on [GSH] after 60 min incubation

Substrates added	[GSH] µmol/g	% change
None	1.46	-33
Serine (2 mM)	1.64	-25
Methionine (0.2 mM)	2.24	+3
Methionine (2 mM)	2.32	+6
Methionine (0.2 mM); serine (2 mM)	2.68	+23
Methionine (2 mM); serine (2 mM)	3.29	+51

Cells from fasted rats prepared in the presence of EGTA. Initial [GSH] 2.17 µmol/g.

Table 4. Effect of homocysteine on the maintenance of [GSH] in the presence of EGTA after 60 min incubation

[Homocysteine] mM	[GSH] µmol/g	% loss
1.0	3.86	13
0.50	3.93	12
0.25	3.67	18
0.12	3.56	20
0.062	3.47	22
0	2.77	38

Initial [GSH] 4.43 µmol/g.

Table 5. Effect of cysteine on [GSH] after 60 min incubation

[Cysteine] mM	[GSH] µmol/g	% GSH left after 60 min
4	0	0
2	0.87	20
1	1.75	41
0.5	2.10	49
0.25	2.24	52
0	3.00	70

Hepatocytes from fed rats; 0.1 mM EGTA.

We thought that a detailed investigation of the cysteine effect may provide clues about the nature of the effects of methionine and homocysteine. We recently found that addition of cysteine to the hepatocytes causes not only a loss of GSH but also of adenine nucleotides, an inhibition of gluconeogenesis and of urea synthesis, a discharge of lactate dehydrogenase and an increased trypan blue uptake. These changes indicate major liver damage. This raises the question of the primary cause of this damage. In order to obtain information leading to the elucidation of the cause of the damage, GSH synthesis was measured in a 15,000 g supernatant of rat liver (and also of rat kidney). The method of Mooz and Meister was used to measured peptide bond synthesis. This is based on the fact that peptide bond synthesis is accompanied by a release of Pi from ATP. The method has the disadvantage of giving a very high blank value under the test conditions, i.e., with impure enzyme solutions. The blank in fact may be three to four times that of the phenomenon under study. Nevertheless, the method gave some clear-cut results.

The rate of Pi release in the presence of glutamate plus glycine represents essentially the blank ATPase activity. Addition of cysteine caused a small inhibition in the liver but a large increase in the kidney (Table 6). When methionine and serine were present in addition to glutamate and glycine, the peptide synthesis was substantially increased in the liver but not in the kidney. Methionine and serine were added because together they cause a gradual release of cysteine, with cystathionine as an intermediate.

In other similar experiments addition of homocysteine plus serine also promoted peptide synthesis. This would be expected if methionine func-

Table 6. Effects of methionine and serine on peptide bond synthesis in high-speed supernatants of rat liver and kidney

Substrate	Liver	Kidney
Glutamate, glycine	0.14	0.08
Glutamate, glycine, cysteine	0.12	0.59
Glutamate, glycine, methionine, serine	0.32	0.13

Peptide bond synthesis was measured by P_i release from ATP according to Mooz and Meister (1974). Initial amino acid concentration was 10 mM. The data are μmol/h/mg protein.

tions as a precursor of cysteine because the combination of homocysteine and serine is nearer to cysteine formation than methionine. The experiments on kidney homogenates show that cysteine is not inhibitory in this tissue. It must be emphasized that it is of importance in experiments with cysteine to neutralise a freshly prepared solution of the hydrochloride immediately before use. Otherwise cysteine solutions may become contamined with cystine. It should also be mentioned that the synthetic capacity of kidney is higher than that of liver (as already well-established by the work of Orlowski and Meister, 1971a,b).

The following ideas emerge about the nature of the methionine and cysteine effects. The toxic effects of cysteine at high concentrations are likely to be connected with the ready oxidation of cysteine to cystine, catalysed either by metal impurities or, perhaps, by the cytochrome P-450 system. The auto-oxidation of cysteine is known to produce not only cystine but also hydrogen peroxide (Harrison and Thurlow, 1926). We suspect that H_2O_2 rather than cystine is the toxic substance on account of its promotion of peroxidative reactions, causing a loss of GSH (see Flohé et al., 1971; Pierce and Tappel, 1978). Whether the loss of GSH is the primary effect, and the loss of adenine nucleotides and intracellular enzymes secondary, is a question which we cannot yet answer.

It is not impossible that cystine formed from cysteine may cause a loss of GSH because the occurrence in liver of a glutathione-cysteine transhydrogenase has been reported by Tietze (1970), though in earlier work Racker (1955) stated specifically that cystine, in contrast to homocystine, does not react as a hydrogen acceptor for the glutathione-disulphide transhydrogenases. However, addition of cystine to isolated hepatocytes caused no loss of GSH. This argues against a ready occurrence of transhydrogenation between GSH and cystine.

The findings suggest that the effects of methionine, or homocysteine, with and without serine, may be due to a gradual release of cysteine needed for the resynthesis of GSH in the course of its rather rapid turnover (Tateishi et al., 1974). As to why liver and kidney differ in major ways, it is relevant that the enzymic equipment of the two organs shows major differences. The enzymes concerned are peroxidases and catalysts responsible for the generation of H_2O_2 such as the cytochrome P-450 system. A gradual release avoids the occurrence of high concentrations of cysteine. Thus it would appear that the methionine effect may not be of major physiological importance because in vivo hepatic cysteine would always arise gradually, either from the blood circulation or from intracellular protein degradation, or possibly from a reduction of cystine. These suggestions are meant to be a working hypothesis guiding further experiments.

References

Berry, M.N., Friend, D.S.: High-yield preparation of isolated rat liver parenchymal cells: a biochemical and fine structural study. J. Cell Biol. 43, 506-520 (1969)
Cornell, N.W., Lund, P., Krebs, H.A.: The effect of lysine on gluconeogenesis from lactate in rat hepatocytes. Biochem. J. 142, 327-337 (1974)
Flohé, L., Eisele, B., Wendel, A.: Glutathion-Peroxidase. I. Reindarstellung und Molekulargewichtsbestimmungen. Hoppe-Seyler's Z. Physiol. Chem. 352, 151-158 (1971)
Harrison, D.C., Thurlow, S.: The secondary oxidation of some substances of physiological interest. Biochem. J. 20, 218-231 (1926)
Krebs, H.A., Cornell, N.W., Lund, P., Hems, R.: Isolated liver cells as experimental material. Alfred Benzon Symp. 6, 726-750 (1974)
Krebs, H.A., Hems, R., Tyler, B.: The regulation of folate and methionine metabolism. Biochem. J. 158, 341-353 (1976)
Mooz, E.D., Meister, A.: Glutathione Biosynthesis. I. γ-Glutamylcysteine synthetase (hog liver). II. Glutathione (tripeptide) synthetase (yeast). Methods in Enzymology 17B, 483-495 (1971)
Orlowski, M., Meister, A.: Isolation of highly purified γ-glutamylcysteine synthetase from rat kidney. Biochemistry 10, 372-380 (1971a)
Orlowski, M., Meister, A.: Partial reactions catalysed by γ-glutamylcysteine synthetase and evidence for an activated glutamate intermediate. J. Biol. Chem. 246, 7095-7105 (1971b)
Pierce, S., Tappel, A.L.: Glutathione peroxidase activities from rat liver. Biochim. Biophys. Acta 523, 27-36 (1978)
Racker, E.: Glutathione-homocystine transhydrogenase. J. Biol. Chem. 217, 867-874 (1955)
Seglen, P.O.: Preparation of rat liver cells. I. Effect of Ca^{2+} on enzymatic dispersion of isolated, perfused liver. Exp. Cell. Res. 74, 450-454 (1972)
Tateishi, N., Higashi, T., Shinya, S., Naruse, A., Sakamoto, Y.: Studies on the regulation of glutathione level in rat liver. J. Biochem. (Tokyo) 75, 93-103 (1974)
Tietze, F.: Disulfide reduction in rat liver. I. Evidence for the presence of nonspecific nucleotide-dependent disulfide reductase and GSH-disulfide transhydrogenase activities in the high-speed supernatant fraction. Arch. Biochem. 138, 177-188 (1970)

The Role of the Cystathionine Pathway in Glutathione Regulation by Isolated Hepatocytes

D. J. REED and P. W. BEATTY

Summary

Freshly isolated hepatocytes were incubated in a complete medium including 10% fetal calf serum and [^{35}S] amino acids. The rate of [^{35}S] GSH synthesis paralleled the rate of uptake of [^{35}S] methionine, [^{35}S] cysteine and [^{35}S] cystine. However, [^{35}S] cystine (0.08 mM) uptake was extremely limited while [^{35}S] cysteine (0.16 mM) uptake was rapid for about 30 min by which time most of the cysteine in the medium had been oxidized to cystine. [^{35}S] methionine (0.67 mM) uptake was about 30% greater than (0.16 mM) cysteine at 30 min and continued at an almost linear rate for the two hour incubations. HPLC analysis permitted quantitation of intracellular and extracellular cysteine, GSH, GSSG and cysteine-glutathione disulfide (CYSSG). GSH biosynthesis was closely related to the efflux of glutathione resulting in GSH, GSSG and CYSSG in the medium. Cystine in the medium stimulated by several fold the presence of CYSSG in the medium at the end of 2 h compared to medium without cystine. It was concluded that the cystathionine pathway is a principal pathway for the maintenance of GSH and that cystine utilization may depend somewhat upon GSH efflux to permit utilization of cysteine derived from a GSH cystine thiol disulfide interchange reaction.

A. Introduction

The biosynthesis of cysteine from methionine via cystathionine was the first pathway for cysteine biosynthesis to be discovered (reviewed by Greenberg, 1975). Further, it appears to be the sole pathway for the formation of cysteine in vertebrates. The role of this pathway in glutathione regulation in liver tissue as well as the consequences of the absence of the pathway in many tumor cell lines is now becoming recognized.

The intracellular level of hepatic cysteine, 0.1-0.3 mM (Tateishi et al., 1974), is much lower than that of glutathione. Much remains to be understood about the regulation and replenishment of the intracellular pools of both thiols including the function of glutathione efflux from liver cells. Tateishi et al. (1974, 1977) have proposed that rat liver glutathione may have a role as a physiological reservoir of cysteine.

Previous work by the author (Reed and Orrenius, 1977) focused upon the question of whether the cystathionine pathway was kinetically important in the synthesis of the thiol reserve present as glutathione in liver cells.

Hepatocytes, freshly isolated from diethylmaleate-treated rats, were shown to perform net biosynthesis of intracellular glutathione at approximately an in vivo rate. This rate of glutathione biosynthesis requires that the total pool of precursor cysteine is replenished every few minutes. The sulfur atom of methionine (Reed and Orrenius, 1977), along with the carbon of serine (Reed and Orrenius, 1978) was found to participate to a major extent in glutathione biosynthesis in glutathione-depleted hepatocytes. Viña et al. (1978) have also suggested that cystathionine may be a cysteine donor for GSH synthesis.

My laboratory has continued a study on the cystathionine pathway and glutathione regulation in hepatocytes. We have confirmed the role of methionine in glutathione biosynthesis under nondepleting conditions. Evidence of a correlation between glutathione efflux and the cyst(e)ine status of the medium has led us to speculate that GSH efflux by hepatocytes may occur in part to facilitate the utilization of cystine via a thiol disulfide interchange, involving GSH and cystine to form cysteine-glutathione disulfide (CYSSG) and cysteine. This process appears to facilitate cystine uptake, which was found to be extremely limited. However, cysteine uptake was extremely rapid; more so than the uptake of methionine. Cysteine oxidation to cystine was a rapid reaction under the incubation conditions demonstrating the necessity for a mechanism for cystine utilization by isolated hepatocytes.

B. Methods

I. Hepatocyte Isolation and Incubation

Hepatocytes were isolated from fed male Sprague-Dawley rats, 200-250 g body wt. as previously described (Reed and Orrenius, 1977) with the following modifications. Calcium was omitted from the collagenase containing perfusion buffer, and the collagenase perfusion time was increased to 15 min. Aliquots of the freshly isolated hepatocytes were counted immediately and hepatocyte suspensions of 10^6 cells/ml were prepared in Fischer's medium (Fischer and Sartorelli, 1964) containing 10% sterile heat-inactivated fetal calf serum. Incubations were at 37°C in 25-ml Erlenmeyer flasks gassed with 95% O_2/5% CO_2 and slowly rotated in a gyratory shaker.

Hepatocyte leakage of lactate dehydrogenase (LDH) was measured by a modified procedure of Lindstrom et al. (1978) with a Beckman TR Analyzer. LDH leakage was measured at hourly intervals with all incubations and found to range from 5% to 15% (except with cystine-containing medium and then hepatocyte leakage increased to about 20% after 2 h).

The procedure was modified from that previously described (Reed and Orrenius, 1977) to permit the formation of S-carboxymethyl rather than sulfonic acid derivatives of thiols. Hepatocytes were removed from incubation mixtures by centrifugation (80 *g* for 2 min) and washed once in saline. The cell pellet was resuspended in saline, 0.05 ml 70% perchloric acid added and the protein removed by centrifugation. A 0.5-ml aliquot of the supernatant was treated immediately with 4 µmol of iodoacetic acid and then neutralized with an excess of $NaHCO_3$.

After 15 min at room temperature, 0.5 ml of an alcoholic solution of 1-fluoro-2,4-dinitrobenzene (FDNB) was added and the reaction allowed to proceed for 4 h in the dark. Aliquots of the perchloric acid supernatant were neutralized and counted in a liquid scintillation counter to determine the acid-soluble ^{35}S fraction whenever [^{35}S] amino acids were utilized in experiments.

II. High Performance Liquid Chromatography (HPLC)

An aliquot of the FDNB reaction mixture was injected onto a 4 x 300 mm Micro Bondapak amine column (Waters) and the DNP derivatives quantitated by gradient elution (0.05-0.4 M sodium acetate, pH 4.6, in 80% methanol). The HPLC system consisted of a Spectra Physics model 3500 liquid chromatograph equipped with a UV detector (350 nm) and a Spectra Physics System I integrator. Appropriate fractions were collected and assayed for radioactivity with an emulsion fluor in a liquid scintillation counter. All derivatives were quantitated by internal standardization with authentic standards.

L-cysteine-glutathione mixed disulfide (CYSSG) was synthesized by thiolysis of the thiolsulfonate derivative of cysteine with glutathione according to the method of Eriksson and Eriksson (1967). The thiolsulfonate derivative of cystine was kindly provided by Dr. Bengt Mannervik (formerly B. Eriksson). All data are the average of three experiments expressed as the mean with standard deviation values.

C. Results

I. Uptake of [^{35}S] Amino Acids

Hepatocytes, 10^6 cells/ml, were incubated in Fischer's medium (Fischer and Sartorelli, 1964) containing 10% fetal calf serum, but lacking in the sulfur amino acids, methionine, cysteine, and cystine except as noted. [^{35}S] Cystine (0.08 mM) uptake was extremely limited, less than 2 nmol/2 h; however, when methionine (0.67 mM) was present in the medium, cystine uptake increased two-fold (Fig. 1).

[^{35}S] cysteine (0.16 mM) uptake was biphasic. Initial uptake was rapid, 15 nmol in 30 min, followed by an uptake rate that was about the rate of [^{35}S] cystine uptake. Cysteine (0.16 mM) in the medium was oxidized to cystine in 30-40 min which could explain the rapid change in uptake rate. In accord with this observation, the presence of methionine

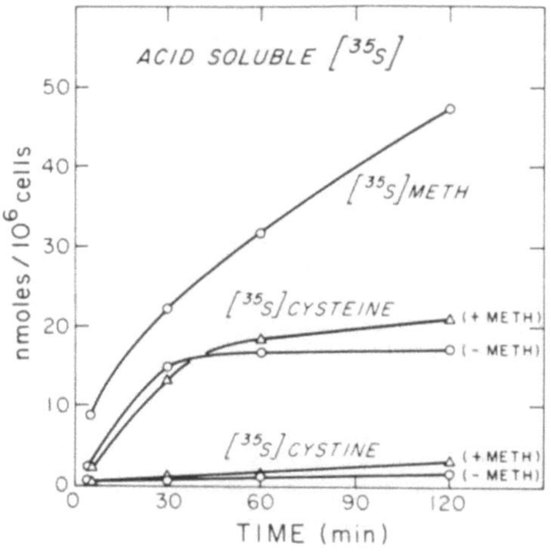

Fig. 1. Uptake of ^{35}S during the incubation of hepatocytes with [^{35}S] cystine (0.08 mM), [^{35}S] cysteine (0.16 mM) or [^{35}S] methionine (0.67 mM). The nmol of acid-soluble ^{35}S were calculated from the specific activity of the [^{35}S] amino acid in Fischer's medium. Sulfur amino acids were deleted and added back as [^{35}S] amino acids except when unlabeled methionine (meth) (0.67 mM) was present as indicated

(0.67 mM) had little effect upon [^{35}S] cysteine uptake for the first hour but caused a significant increase during the second hour. [^{35}S] methionine (0.67 mM) uptake was nearly linear and resulted in a total uptake of 45 nmol/10^6 cells/2 h.

II. Specific Activity of Intracellular [^{35}S] GSH

The rate of increase in [^{35}S] GSH specific activity paralleled the rate and extent of uptake of the [^{35}S] amino acids (Fig. 1, Table 1). The specific activity increase of [^{35}S] cystine-derived [^{35}S] GSH was linear as was the uptake rate of [^{35}S] cystine from the medium. [^{35}S] GSH labeling from [^{35}S] cysteine occurred rapidly during the first hour and then increased at a much slower rate, about twice the rate of labeling by [^{35}S] cystine during the second hour. [^{35}S] GSH specific activity was 24% of that of [^{35}S] cysteine in the medium after 2 h representing about a 6-fold greater labeling than from [^{35}S] cystine.

[^{35}S] methionine uptake rate was greater than [^{35}S] cysteine and also gave greater labeling of GSH (Fig. 1, Table 1). At 2 h, after an almost linear rate of increased labeling, [^{35}S] GSH specific activity was 36% of the specific activity of [^{35}S] methionine in the medium. It should be noted that at the 1-h interval methionine achieved the same degree of ^{35}S labeling of GSH as cysteine, yet [^{35}S] methionine was at a four-fold higher concentration than cysteine in the medium. Also, total [^{35}S] methionine uptake was twice [^{35}S] cysteine uptake reflecting in part a greater intracellular pool of acid-soluble [^{35}S]. These methionine intermediates may represent in part [^{35}S] labeled intermediates of the cystathionine pathway.

III. Intracellular GSH Content During [^{35}S] GSH Formation

None of the three amino acids maintained intracellular GSH content of the hepatocytes at the initial level (Table 2). The greatest decrease

Table 1. [^{35}S] Amino acid incorporation into intracellular GSH of isolated hepatocytes

[^{35}S] Amino acid	Conc. (mM)	[^{35}S] GSH specific activity expressed as percentage of ^{35}S amino acid specific activity[a]		
		Time (min)		
		10	60	120
Cystine	0.08	0.1 ± 0.1	2.2 ± 0.4	4.2 ± 0.2
Cysteine	0.16	1.7 ± 1.0	20.2 ± 2.2	24.1 ± 1.0
Methionine	0.67	2.0 ± 2.1	19.3 ± 1.4	36.4 ± 2.2

[a] [^{35}S] cystine specific activity adjusted to cysteine equivalents.

Table 2. Intracellular GSH content during [^{35}S] amino acid incorporation into isolated hepatocytes

Amino acid	Conc. (mM)	GSH content (n mol/10^6 cells)		
		Time (min)		
		10	60	120
Cystine	0.08	68.8 ± 13.3	50.0 ± 7.7	33.9 ± 1.4
Cysteine	0.16	50.2 ± 13.7	49.6 ± 15.3	41.4 ± 12.9
Methionine	0.67	53.5 ± 11.5	42.9 ± 8.5	43.7 ± 18.8

occurred with cystine in the medium (a decrease of 35 nmol/10^6 cells) while cysteine and methionine containing media resulted in the cellular GSH being 40-44 nmol/10^6 cells after 2 h of incubation. From the total ^{35}S uptake, the specific activity of [^{35}S] GSH, and the intracellular content of GSH, it is possible to calculate the percent of total acid-soluble ^{35}S that is present as GSH in the hepatocytes. At the end of 2 h, [^{35}S] GSH, derived from [^{35}S] cystine or ^{35}S cysteine, was nearly 60% of the total acid soluble [^{35}S]. [^{35}S] GSH derived from [^{35}S] methionine was 40% of the total.

IV. Extracellular Glutathione

Sensitivity of the HPLC technique employed in these experiments permitted the quantitation of not only GSH and GSSG but also cysteine-glutathione mixed disulfide (CYSSG). We conducted the initial sampling of the hepatocytes as rapidly as possible after placing them in media, but it still required 6 to 9 min because of the initial sedimentation by centrifugation (80 g for 2 min), and washing the pelleted hepatocytes with saline. However, in every instance an appreciable extracellular level of GSH, 3-5 nmol/10^6 cells, was present in the medium (Table 3). During the next 2 h this level increased from 6 to 7 nmol, slightly higher than the GSSG levels at 2 h. Thus, a GSH/GSSG ratio of about 1 was observed. In contrast, CYSSG levels were much higher in media containing cysteine and cystine, 19-22 nmol/10^6 cells at the 2-h interval, than when the medium contained methionine. A much lower level of CYSSG was present with the methionine-containing medium and, presumably, both the cysteine and GSH moiety of CYSSG originated by efflux from hepatocytes.

V. Total Glutathione Status

In an effort to understand more fully the kinetics of GSH biosynthesis from precursor amino acids by isolated hepatocytes, the status of total acid-soluble glutathione was considered in terms of both intracellular and extracellular glutathione and CYSSG. For example, while extracellular CYSSG increased 19 nmol during the 120-min period of incubation of hepatocytes in cystine-containing medium, intracellular GSH decreased 35 nmol, extracellular GSH and GSSG increased 10 nmol leaving 6 nmol unaccounted for (Table 4). CYSSG in the medium represented 66% of the total glutathione efflux. Protein glutathione mixed disulfide content of cells or medium was not measured in these experiments. Total glutathione change represented increases of 12% and 39% when hepatocytes were incubated in media containing either methionine or cysteine respectively. These increases were accompanied by the appearance of CYSSG in the media.

Table 3. Extracellular GSH, GSSG and CYSSG levels during [^{35}S] amino acid incorporation into isolated hepatocytes

[^{35}S] Amino acid	Conc. (mM)	GSH		GSSG (nmol/10^6 cells)		CYSSG	
		Time (min)					
		10	120	10	120	10	120
Cystine	0.08	5.4 ± 0.5	6.5 ± 1.8	1.0 ± 0.3	5.6 ± 3.1	2.9 ± 0.1	22.3 ± 5.2
Cysteine	0.16	3.1 ± 1.0	6.7 ± 1.3	<0.3	5.5 ± 2.6	2.6 ± 0.3	19.1 ± 2.0
Methionine	0.67	3.3 ± 0.1	6.6 ± 2.4	<0.3	4.7 ± 1.6	1.7 ± 0.4	6.4 ± 1.3

Table 4. Effect of sulfur amino acids in the medium on total glutathione status

Amino acid	Conc. (mM)	Total glutathione increase/2 h (nmol/10^6 cells)	% increase	% increase as extracellular CYSSG
Cystine	0.08	-5.2	-7	–
Cysteine	0.16	22.0	39	87
Methionine	0.67	7.3	12	88

D. Discussion

Evidence has been presented which supports the concept of an important role for extracellular sulfur amino acids as precursors for glutathione biosynthesis by isolated hepatocytes. Their relative roles as precursors appear to be closely related to the rate of uptake of each amino aicd. Cysteine uptake appears to be at least twice as rapid as that of methionine while cystine has a very slow rate of uptake.

The almost linear rate of [^{35}S] methionine uptake and the high degree of conversion of ^{35}S to [^{35}S] GSH strongly support the concept of a very active role for the cystathionine pathway in the biosynthesis and regulation of GSH. With all the [^{35}S] amino acids examined it can be calculated that the degree of GSH labeling is in reasonably good agreement with the amount of biosynthesis required on the basis of total glutathione efflux which includes the presence of CYSSG in the medium of incubating hepatocytes. From the data presented here it is not possible to evaluate the degree to which GSH undergoes cellular degradation and resynthesis as indicated in Figure 2.

Rapid uptake of cysteine may related to the fact that a majority of the cyst(e)ine in plasma is cystine (Crawhall and Segal, 1967) and therefore cysteine uptake must be very efficient. In addition, one may expect that mechanisms may exist for cystine utilization via the extracellular conversion of cystine to cysteine. Some in vivo evidence exists for such mechanisms.

Schneider et al. (1968) reported that a human suffering from homocystinuria had plasma levels of cysteine-homocysteine disulfide and methionine 12-fold higher than control humans, while the cystine level was only 12% of normal. Crawhall and Segal (1967) found cyst(e)ine in rat tissues mainly in the reduced form, while plasma cyst(e)ine was mainly as cystine. They also noted that under anaerobic conditions tissue slice incubations resulted in cystine in the medium becoming reduced. They speculated that cysteine and/or glutathione were effluxing from the tissue and liberating cysteine from cystine in the medium by thiol-disulfide interchange. They further speculated that this phenomenon was occurring in aerobic medium also, but that it could be obscured because of rapid reoxidation of the cysteine. Could the efflux of GSH from liver play a similar role in cystine utilization?

Assuming 1.2 x 10^8 hepatocytes/g wet wt. of liver, Bartoli and Sies (1978), using a hemoglobin-free perfusing medium, observed a combined GSH and GSSG efflux of 7.2 nmol/h/10^6 cells with liver perfusion experiments. Over 90% of the glutathione efflux was GSH. Previously Reed and Orrenius (1977) using a medium containing either cysteine or methionine reported total glutathione efflux (as measured by a performic oxidation procedure) of up to 7-8 nmol/h/10^6 cells.

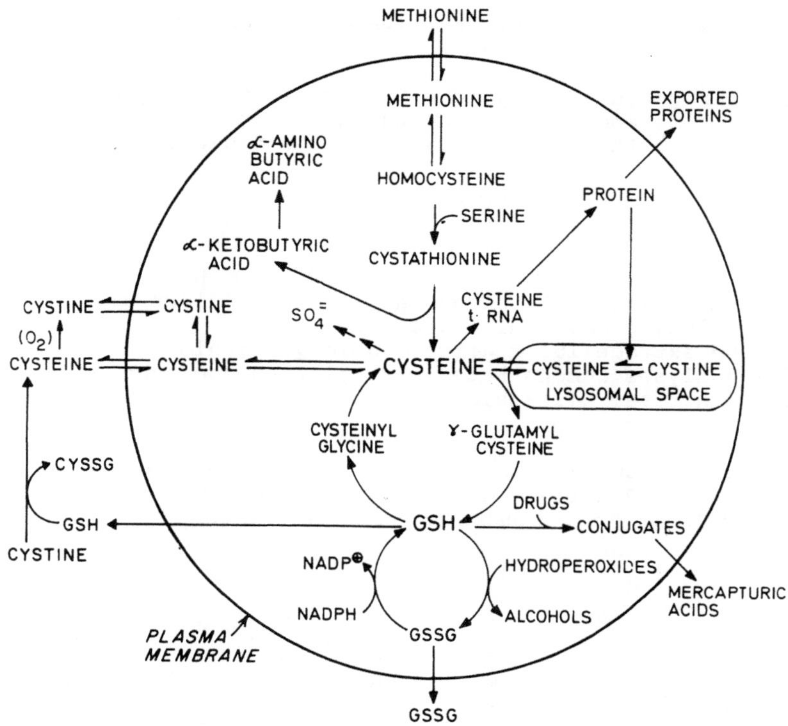

Fig. 2. The possible inter-relationships of the sulfur amino acids for glutathione regulation in hepatocytes

The presence of GSSG and GSH in the medium within 10 min after adding a hepatocyte suspension was not due to pre-existing acid soluble thiols in the medium since analysis of Fischer's medium containing 10% fetal calf serum showed no detectable GSH, GSSG or CYSSG. Fetal calf serum is unique among many sera in that it contains 10-23 nmol of glutathione/ml (Bump and Reed, 1977) as protein-mixed disulfides. However, if the entire amount became available it could contribute less than 3 nmol/10^6 cells at the 10% concentration of serum used in the medium.

One can speculate that glutathione efflux by hepatocytes may actually represent a glutathione gradient of intra- and extracellular GSH. If so, in these experiments the gradient would be established at approximately 5 nmol/ml in the medium with 40 to 50 nmol/10^6 cells/ml medium. Assuming a hepatocyte volume of 7.25 µl/10^6 cells (Krebs et al., 1974) then the concentration gradient of the GSH content of cells versus medium is nearly 1400 to 1. Tietze (1969) observed a plasma glutathione concentration of 5 nmol/ml in blood samples obtained by heart puncture of rats. The presence of cystine in the medium could lead to further GSH efflux as thiol disulfide interchange occurred (Fig. 2). The rate at which this occurred would depend on the rate of interchange. Presumably some CYSSG-thioltransferase enzyme (Eriksson et al., 1974a and 1974b) could be released into the medium from hepatocytes.

The thiol-disulfide interchange reaction between cystine and GSH:

$$\text{GSH} + \text{cystine} \rightleftarrows \text{CYSSG} + \text{cysteine}$$

has an equilibrium constant of 3.2 at pH 7.4 and 37° (Jocelyn, 1967) which favors the formation of cysteine.

Finally one can further speculate that the GSH gradient, being a membrane process, might be dependent upon the thiol content of the medium and thus responsive to the thiol redox status of the medium.

E. Conclusions

1. Hepatocytes can synthesize glutathione rapidly from sulfur amino acid precursors including methionine and cysteine.
2. Combined intra- and extracellular acid-soluble glutathione (GSH, GSSG, and CYSSG) analyses demonstrated that glutathione synthesis correlated well with total glutathione efflux.

Acknowledgments. This work was supported in part by the U.S. National Institute of Health Grant ES-01978-01 and IF32 ES 05087-01 (P.W.B.). The authors wish to thank Dr. Ann Brodie and Joanne Potter for invaluable technical assistance.

References

Bartoli, G.M., Sies, H.: Reduced and oxidized glutathione efflux from liver. FEBS Lett. 86, 89-91 (1978)

Bump, E.A., Reed, D.J.: A unique property of fetal bovine serum: High levels of protein-glutathione mixed disulfides. In Vitro 13, 115-118 (1977)

Crawhall, J.C., Segal, S.: The intracellular ratio of cysteine and cystine in various tissues. Biochem. J. 105, 891-896 (1967)

Eriksson, B., Eriksson, S.A.: Synthesis and characterization of the L-cysteine-glutathione mixed disulfide. Acta Chem. Scand. 21, 1304-1312 (1967)

Eriksson, S., Askelöf, P., Axelsson, K., Mannervik, B.: Evidence for a glutathione-dependent interconversion of two forms of a thioltransferase from rat liver catalyzing thiol-disulfide interchange. Acta Chem. Scand. B28, 931-936 (1974a)

Eriksson, S., Askelöf, P., Axelsson, K., Carlberg, I., Guthenberg, C., Mannervik, B.: Resolution of glutathione-linked enzymes in rat liver and evaluation of their contribution to disulfide reduction via thiol-disulfide interchange. Acta Chem. Scand. B28, 931-936 (1974b)

Fischer, G.A., Sartorelli, A.C.: Development, maintenance and assay of drug resistance. In: Methods in medical research 10, 247-262 (1964)

Greenberg, D.M.: Biosynthesis of cysteine and cystine. In: Metabolic pathways, Vol. VII: Metabolism of sulfur compounds. Green, D.M. (ed.) pp. 505-528, Academic Press: New York 1975

Higashi, T., Tateishi, N., Naruse, A., Sakamoto, Y.: A novel physiological role of liver glutathione as a reservoir of L-cysteine. J. Biochem. 82, 117-124 (1977)

Jocelyn, P.C.: The standard redox potential of cysteine-cystine from the thiol-disulfide exchange reaction with glutathione and lipoic acid. Eur. J. Biochem. 2, 327-331 (1967)

Krebs, H.A., Cornell, N.W., Lund, P., Hems, R.: Isolated liver cells as experimental material. Alfred Benzon Symp. 6, 718-743 (1974)

Lindstrom, T.D., Anders, M.W., Remmer, H.: Effect of phenobarbital and diethyl maleate on carbon tetrachloride toxicity in isolated rat hepatocytes. Exp. Mol. Pathol. 28, 48-57 (1978)

Reed, D.J., Orrenius, S.: The role of methionine in glutathione biosynthesis by isolated hepatocytes. Biochem. Biophys. Res. Commun. 77, 1257-1264 (1977)

Reed, D.J., Orrenius, S.: The protective role of methionine and glutathione biosynthesis during drug metabolism by isolated hepatocytes. Toxicol Appl. Pharmacol., 45, 241 (1978)

Schneider, J.A., Bradley, K.H., Seegmiller, J.E.: Identification and measurement of cysteine-homocysteine mixed disulfide in plasma. J. Lab. Clin. Med. 71, 122-125 (1968)

Tateishi, N., Higashi, T., Shinya, S., Naruse, A., Sakamoto, Y.: Studies on the regulation of glutathione level in rat liver. J. Biochem. 75, 93-103 (1974)

Tateishi, A., Higashi, T., Naruse, A., Nakashima, K., Shiozaki, H., Sakamoto, Y.: Rat liver glutathione: Possible role as a reservoir of cysteine. J. Nutrition 107, 51-60 (1977)

Viña, J., Hems, R., Krebs, H.A.: Maintenance of glutathione content in isolated hepatocytes. Biochem. J. 170, 627-630 (1978)

Physiological Significance of Elevation of Liver γ-Glutamyltranspeptidase in Late Pregnancy

T. HIGASHI and N. TATEISHI

Summary

γ-Glutamyltransferase (EC 2.3.2.2) activity in mouse fetal liver increases in late pregnancy to maximum (300- to 500-fold that of adult liver) at birth and then decreases to nearly the level in adult liver within one week. Glutathione synthesis steadily increases during the perinatal period. Liver glutathione increases during the fetal period, but reaches less than half the adult level. It decreases transiently just before birth, and then increases after birth. The alteration in glutathione appears to reflect the activities of these enzymes involved in glutathione synthesis and degradation.

In studies on the physiological meaning of the transient increase of γ-glutamyltransferase, 6-diazo-5-oxo-L-norleucine was used to inhibit the enzyme activity in vivo. On injection of this substance into pregnant mice at the end of pregnancy, the activity of fetal liver γ-glutamyltransferase was strongly inhibited with concomitant increase in liver glutathione. Tracer studies with [^{35}S]cysteine showed increased radioactivity in glutathione and decreased incorporation into liver cysteine in the presence of the inhibitor. In the absence of the inhibitor, [^{35}S]cysteine transferred from the mother to the fetuses was mainly incorporated into liver proteins, with some incorporation into glutathione, especially on the last day of gestation. The enzyme in the mother's liver also increased in this period. Increased γ-glutamyltransferase in both the mother and pups probably ensures an adequate supply of cysteine to meet the increasing demands of the pups for a proper level of cysteine.

A. Introduction

Since Dr. Meister and his colleagues proposed the "γ-glutamyl cycle", in which γ-glutamyltranspeptidase plays a central role, there has been much debate about the physiological function of this enzyme in the plasma membrane. Little attention has been paid to γ-glutamyltranspeptidase (γ-GTP) in the liver, because its activity is much lower than that in the kidney, although it has been shown to participate in the hepatic detoxication mechanism.

We observed an increase of liver γ-GTP under certain conditions, such as prolonged starvation, chronic diabetes and late pregnancy.

B. Elevation of Liver γ-GTP in the Latest Part of Pregnancy

Of all the conditions studied, late pregnancy was found to cause the greatest induction of γ-GTP in the liver. The γ-GTP activity in mouse fetal liver was greatly elevated in late pregnancy, reaching a maximum at birth and then decreasing rapidly within one week after birth to nearly the level in adult liver. The enzyme activity at birth was 300-500 fold that of adult mouse liver (1). Similar induction of γ-GTP was found in perinatal rat liver. This temporary but significant increase of γ-GTP around birth seemed worth analysis in order to elucidate the role of the enzyme in nutritional adaptation of the animals during this critical period of life.

C. Hepatic Levels of Glutathione and Cysteine in Developing Mice

To examine whether this transient increase of γ-GTP was directly related to the metabolism of glutathione, we measured the hepatic levels of glutathione and cysteine in developing mice. The level of glutathione rose in fetal liver during development, but it reached less than half the adult level and decreased immediately before birth. It then steadily increased again after birth. The level of cysteine increased just before birth and decreased gradually after birth, although its apparent change was much less than that of glutathione. The transient pause of the increase in glutathione coincided with the period when γ-GTP activity was maintained at a high level (Table 1).

D. Increase in Activities of Glutathione Synthesizing Enzymes in Developing Mouse Liver

The enzymes other than γ-GTP directly involved in glutathione metabolism are the glutathione-synthesizing enzymes, γ-glutamylcysteine synthetase and glutathione synthetase. Their overall activity, estimated by measuring incorporation of ^{14}C-glycine into glutathione in the cytosol

Table 1. Changes of hepatic glutathione and cysteine levels and of γ-glutamyltranspeptidase and glutathione-synthesizing enzyme activities in developing mice

		GSH[a]	Cysteine[b]	γ-GTP[c]	Glutathione-synthesizing enzyme[d]
Day of	13	-[e]	-	1.9	-
gestation	14	-	-	-	35
	15	2.6	0.23	35	-
	16	3.5	0.25	51	38
	17	3.65	0.27	-	42
	18	3.5	0.33	129	-
	19	3.3	0.47	185	50
Day after	0	2.94	0.55	207	52
birth	1	3.45	0.49	176	-
	2	-	-	138	-
	3	4.4	0.48	111	70
	4	4.95	0.37	41	85
	5	-	-	-	100
	6	-	-	-	-
	7	5.55	0.33	5.9	117

[a]µmol/g liver; [b]µmol/g liver; [c]µmol of γ-glutamyl-p-nitroanilide hydrolyzed/h/g liver at 25°C; [d]nmol of ^{14}C-glycine incorporated into glutathione/min/ml cytosol at 37°C; [e]not determined.

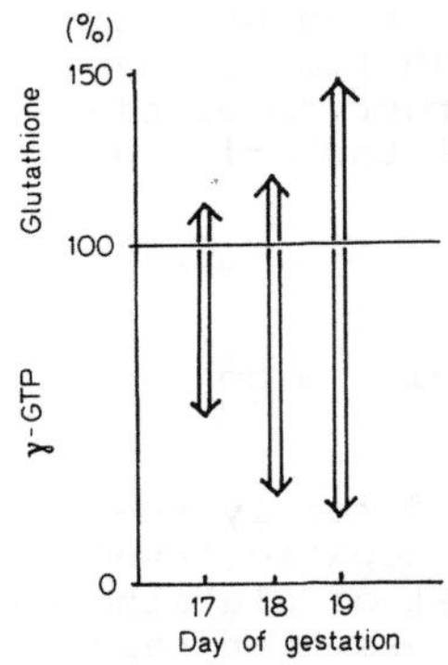

Fig. 1

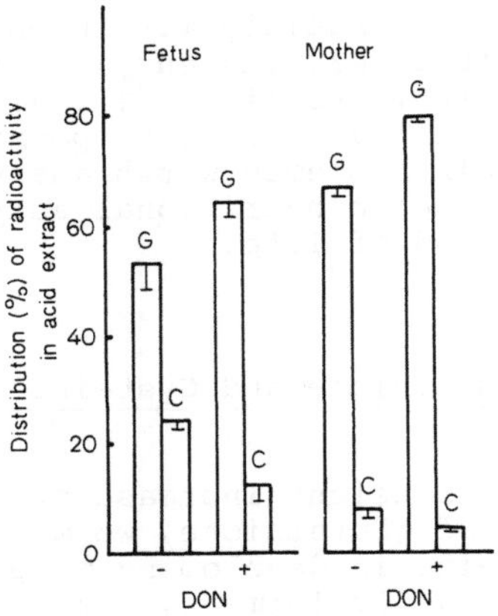

Fig. 2

Fig. 1. Inactivation of fetal hepatic γ-glutamyltranspeptidase and accumulation of glutathione in fetal liver by treatment with DON. DON (25 μmol) or saline was injected to pregnant mice intravenously on gestational day 17, 18, or 19 and the animals were killed 4 h after injection. γ-glutamyltranspeptidase activity and glutathione concentration were examined in fetal liver. Three to five mothers were used for each experiment. The figures for saline-injected group was taken as 100%

Fig. 2. The effect of DON on the incorporation of L-[^{35}S]-cysteine into liver glutathione and cysteine. DON (25 μmol) was injected to the pregnant mice on gestational day 19 as described in the legend of Fig. 1. One hour later, 6 μCi/0.25 ml saline of L-[^{35}S]-cysteine was injected into tail vein under light ether anesthesia. Three hours later, the animals were killed. In the acid extract of liver (total radioactivity = 100%), glutathione and cysteine were separated by paper electrophoresis and thin layer chromatography of hepatic acid extract. Most of the residual radioactivity in the acid extract was found in the taurine fraction. *Vertical bars* in the columns indicate standard deviations of the values from 3 to 4 mothers. C, cysteine (cystine); G, glutathione

fraction of the liver, increased throughout the perinatal period (Table 1) (2).

If γ-GTP and glutathione synthesizing enzyme cause the observed changes in glutathione and cysteine, it should be possible to observe changes in the hepatic levels of glutathione and cysteine by selectively inhibiting the activity of γ-GTP. Therefore, we tested the effect of 6-diazo-5-oxo-L-norleucine (DON), which is a selective inhibitor of this enzyme.

E. Accumulation of Hepatic Glutathione in the Presence of DON

Injection of 25 μmol of DON into a tail vein of pregnant mice on day 19 of pregnancy resulted in severe and rapid inactivation of γ-GTP in the fetal organs, with less inactivation of γ-GTP in the mother's kidney. The inhibitions of fetal γ-GTP on days 17 and 18 were less than that on day 19 (Fig. 1). Concomitantly, injection of DON prevented the transient decrease of liver glutathione level at the end of fetal life. On day 19, the glutathione level in fetal liver in the presence of DON was 50% higher than that in control fetuses (Fig. 1)(3). Intraperitoneal injection of ^{35}S-cysteine into pregnant mice treated with DON resulted in increased labeling of glutathione and decreased labeling of cysteine in the fetal liver. Similar changes in the distribution of radioactivity were observed in the mother's liver (Fig. 2)(3). These findings suggest that under normal conditions, cysteine once incorporated into glutathione must have been released readily by the action of the increased γ-GTP at the fetal period, leading to the decrease of glutathione in the liver.

F. Significance of Increased γ-GTP at the End of Pregnancy

We designed an experiment to determine the physiological significance of the abrupt change of γ-GTP activity in the perinatal period. Doses of 4 μCi of ^{35}S-cysteine were injected into a tail vein of mice on days 15-19 of pregnancy and in the lactating period. The animals were killed 3 h after the injections, and the distribution of radioactivity in their liver was examined. The ratio of radioactivity in whole liver protein in the pups to that in the mothers showed that at the end of pregnancy cysteine was transferred preferentially from the mother to fetus for protein synthesis (Fig. 3). Fetal protein and glutathione fractions always showed higher specific radioactivity than the corresponding mother fractions. At the same time, cysteine was mobilized from glutathione in the fetal liver for the same purpose. The γ-GTP, which increased to a maximum at birth, must be responsible for the mobilization of cysteine from liver glutathione. A similar increase in turnover of liver glutathione in the mother's liver is suggested by the increase of hepatic γ-GTP activity in pregnant animals. All these changes seem to meet the increased demand of the pups for the amino acid. The fetal system for cysteine release continued to work for a while after birth, when the placenta no longer supplied cysteine and the milk obtained by suckling was not yet sufficient for the pups. Glutathione synthesizing enzyme activity in the liver of pups continued to increase in this period. The increase of glutathione synthesis together with the high activity of γ-GTP appears to form a "futile" cycle of glutathione and cysteine. This is, however, a device for maintaining the concentration of cysteine at a harmlessly low level and ensuring an adequate supply of cysteine, even when the requirement for

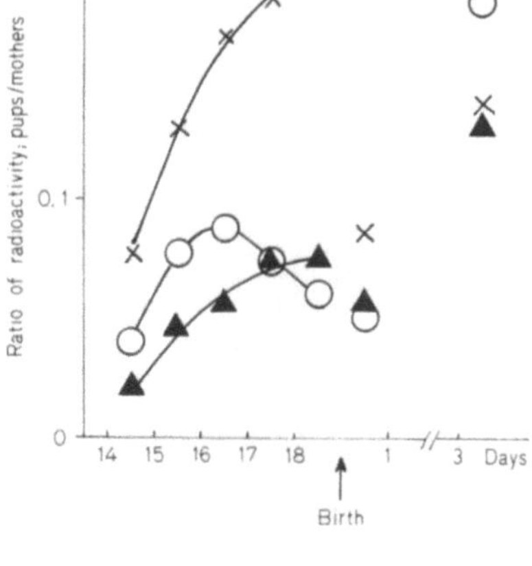

Fig. 3. Transfer of ^{35}S-cysteine from mothers to offsprings via placenta or milk. Doses of 4 μCi of L-[^{35}S]-cysteine were injected into a tail vein of pregnant mice on gestational day 15 to 19 and lactating mothers. The animals were killed 3 h later, the mothers' livers were dropped into liquid nitrogen as quickly as possible and the fetuses were removed rapidly by Caesarean section. The livers of all the fetuses from one mother were pooled in liquid nitrogen. The distribution of radioactivity in the liver of mother and pups was examined. The *ordinate* indicated the ratio of total radioactivity in a given fraction of whole liver of pups to that of mothers' liver. The *abscissa* indicates the days of gestation and days after birth. L-[^{35}S]-cysteine was transferred transplacentally before birth and via milk after birth. Two to five mothers were used for each experiment. X, proteins; O, glutathione; ▲, "cysteine"

cysteine is maximally increased. γ-GTP in the liver appears to be the enzyme responsible for ensuring the adequate supply of cysteine.

References

1. Tateishi, N., Higashi, T., Nomura, T., Naruse, A., Nakashima, K., Shiozaki, H., Sakamoto, Y.: Higher transpeptidation activity and broad acceptor specificity of γ-glutamyltransferases of tumors. Gann 67, 215-222 (1976)
2. Tateishi, N., Higashi, T., Nakashima, K., Sakamoto, Y.: Nutritional significance of increase in mouse liver γ-glutamyltransferase in late pregnancy. (Submitted for publication)
3. Tateishi, N., Higashi, T., Sakamoto, Y.: Accumulation of hepatic glutathione in mouse late fetal life after treatment with 6-diazo-5-oxo-L-norleucine (DON). (Manuscript in preparation)

Glutathione Efflux from Perfused Rat Liver and Its Relation to Glutathione Uptake by the Kidney

G. M. BARTOLI, D. HÄBERLE, and H. SIES

Summary

The isolated perfused rat liver releases GSH at a rate of 12 nmol/min/g wet wt. at 37°C, whereas the release of GSSG is much lower, 1 nmol/min/g. It has recently become evident that glutathione release from liver may be of significance for glutathione turnover if viewed in conjunction with transport in plasma and degradation by extrahepatic tissues, notably the kidney. We have found that GSH efflux occurs into the caval perfusate, while GSSG efflux occurs into bile, so that GSH efflux into plasma may also occur in vivo. Assuming that GSH release from liver occurs at a rate of 12 nmol/min/g liver also in the intact animal, it is calculated that the glutathione pool designated as the "labile pool" with a half-life time of 1.7 h would have a pool size of 1.8 μmol/g liver, amounting to approximately one-third of the glutathione content. If, on the other hand, the total glutathione pool would contribute to the efflux, the half-life time would have to be approximately 5.2 h.

In order to estimate the capacity of renal plasma glutathione degradation, arterial and renal venous plasma glutathione concentrations were determined in samples obtained from anaesthetized rats. The glutathione concentration in arterial plasma was 3.1 μM (expressed in GSH equivalents), and the renal venous plasma value was 88% lower. This indicates that glutathione uptake by the kidney occurs not only by glomerular filtration and subsequent degradation by the γ-glutamyltranspeptidase of the tubular brush border membrane, but also by an additional uptake mechanism. The rate of glutathione uptake is limited by the renal plasma flow, since the glutathione extraction (%) was similar to the controls in rats receiving a GSH infusion to result in arterial plasma values up to 0.6 mM.

It is concluded from these data that the GSH release observed in the perfused liver is compatible with the in vivo half-life time, and that the kidney has the capacity to extract a significant part of the plasma glutathione. However, a calculation with the renal plasma flow rates indicates that organs other than the kidney also play a role in plasma glutathione turnover.

A. Introduction

It was early recognized that the capacity for enzymatic hydrolysis of glutathione is localized mainly in the kidney (1). More recently, evidence for interorgan relationships in the turnover of glutathione has accumulated (2-5). Thus, the concept has emerged of the kidney as a major site of glutathione degradation to the constituent amino acids, whereas other organs, e.g., the liver, synthesize their onw glutathione pool and contribute to the plasma pool in an as yet ill-defined manner. Tietze (6) has detected a low steady-state glutathione concentration in rat plasma.

Recently, we have described the release of glutathione as GSH from the isolated perfused rat liver (7). The observed rate was compatible with the in vivo turnover time of liver glutathione. Therefore, it was of further interest to estimate the capacity of renal plasma glutathione degradation in the intact organism. For this purpose, we determined glutathione in samples of arterial and renal venous plasma obtained from anaesthetized rats under steady-state conditions.

B. Materials and Methods

I. Hemoglobin-free Perfusion of Rat Liver

Male Wistar rats of 140-180 g of body wt., fed on stock diet (Altromin) were perfused at 37°C as described (8). Perfusate flow rate was 4 ml/min/g liver. Effluent perfusate was assayed directly for lactate dehydrogenase activity, for glutathione disulfide (GSSG) (9), glutathione (GSH) (10), and GSH and 2 GSSG in the catalytic assay (6,11).

II. Glutathione in Arterial and Renal Venous Plasma

Male Wistar rats of 350-420 g of body wt., fed on stock diet (Altromin 1320), were anaesthetized with the thiobarbiturate, Inactin (Byk Gulden, Konstanz), and further prepared as described (12,13). Blood pressure was monitored on the right femoral artery throughout the experiment. A solution of 4 g of polyfructosan (Inutest, Laevosan, Linz) per 100 ml of isotonic saline was infused into the animals' right jugular vein at a rate of 1.5 ml/h/100g body wt. Further details of the sampling described later are given in the legend of Table 2. Perchloric acid extracts of plasma samples were neutralized directly before the catalytic assay for glutathione (6,11) which measured both GSH and GSSG; data are expressed in GSH equivalents. Standardized glutathione was added to each sample, since in neutralized extracts the reaction velocity obtained by standard addition was approximately two-thirds of comparable blank extracts.

C. Results and Discussion

I. Rates of Glutathione Release from Perfused Rat Liver

The release of glutathione into the effluent perfusate was measured by three different methods. As shown in Table 1 (7), GSSG is released at a rate of 1 nmol/min/g liver, whereas GSH is released at a rate of 11.8 nmol/min/g liver, so the total amounts to 13.8 nmol/min/g as expressed in GSH equivalents. The value found with the third method, using Ellman's reagent and glutathione reductase, is 14.5 nmol/min/g, again expressed as GSH and 2 GSSG. As shown elsewhere (14,15), the GSSG release occurs into bile, whereas GSH release occurs into the perfusate. In agreement with earlier observations (9,16), the extra release of glutathione elicited by the addition of t-butyl hydroperoxide to the influent perfusate occurred in the form of GSSG (Table 1), leaving the release of GSH unaltered.

The total GSH content of the liver is 5.28 µmol/g (16), so that the observed rate of release corresponds to 0.3% of total glutathione/min.

Table 1. Release of glutathione and of lactate dehydrogenase from hemoglobin-free perfused rat liver

Parameter	Rate of release from liver	
	Control	t-Butyl hydroperoxide (0.5 μmol/min/g)
GSH (nmol/min/g)	11.8 ± 0.4(8)	11.2
GSSG	1.0 ± 0.1(8)	18.7
GSH + 2GSSG,in GSH equiv.	13.8	48.8
GSH + 2GSSG,in GSH equiv. (determined with catalytic assay)	14.5 ± 0.8(8)	45.2
Lactate dehydrogenase activity (mU/min/g)	11.2 ± 1.6(5)	12.0

Perfusate flow, 4 ml/min/g liver wet wt. temp. 37°C. Data are expresses as means ±s.e. (no. different perfusions in parentheses). Values were determined in triplicate in each experiment (three for addition of t-butyl hydroperoxide) (7).

Thus, the observed release of glutathione is compatible with the reported half-life times for hepatic glutathione (17-19). It is possible, therefore, that the observed release of glutathione from the perfused liver reflects a physiological process rather than an experimentally induced leakage. This is supported by the low rate of release of lactate dehydrogenase (Table 1) which corresponds to 0.004% of total lactate dehydrogenase/min, i.e., to a rate of two orders of magnitude lower than that of glutathione.

Assuming that the GSH release from liver occurs at a rate of 12 nmol/min/g liver also in the intact animal, it is calculated that the glutathione pool designated as the "labile pool" with a half-life time of 1.7 h (17) would have a pool size of 1.77 μmol/g liver, amounting to approximately one-third of the glutathione content. If, on the other hand, the total glutathione pool would contribute to the efflux (which is considered less likely), the half-life time would have to be approximately 5.2 h.

II. Glutathione Uptake by the Kidney in the Anesthetized Rat

The physiological significance of the observed glutathione efflux from the perfused liver remains to be evaluated. With a liver weight of 4.35 g/100 g body wt., the observed rate of 12 nmol GSH/min/g liver would correspond to 52 nmol/min/100 g of rat if the efflux occurs at a similar rate in vivo. This has not been tested so far.

However, as the kidney has been identified as a major site of degradation of extracellular glutathione (3-5), we have carried out the following measurements on the capacity of renal plasma glutathione degradation in the intact organism (to be published). The arterial and renal venous plasma glutathione concentrations were determined at control conditions and during the infusion of GSH (Table 2). The control value of 3.1±0.4 (n=7) μM glutathione for arterial plasma is slightly lower than the value of about 5 μM (no statistical information available) reported earlier (6). Interestingly, the mean renal venous glutathione concentration is only 12.5% of the arterial value, i.e., the uptake by the kidney is 87.5%. Since the glomerular filtrate amounted to only 19.3% of the renal plasma flow, this means that glutathione degradation occurs not only by glomerular filtration and subsequent degradation by the γ-glutamyltranspeptidase of the tubular brush border membrane sim-

Table 2. Glutathione concentration in arterial and renal venous plasma of the rat

Sample	Glutathione as GSH+2GSSG(μM) arterial	renal venous	Glutathione extraction %	Renal plasma flow ml/g kidney/min	Filtration fraction %
Control	3.1±0.4(7)	0.39±0.08(7)	87.5±2.5(7)	2.88±0.81(3)	19.3±3.9(7)
GSH infusion	667±112(4)	160 ±54.6(4)	79.2±6.4(4)	3.35±0.40(4)	32.5±2.2(4)

Relevant details of the experimental protocol (to be given in detail elsewhere) are the following: After an equilibration period of 45 min after completion of the surgery, the animals were heparinized (150 U.S.P.-E Heparin/kg), and urine from both kidneys was collected over further 45 min; simultaneously, 1.5 ml of blood was taken from the femoral artery. 15 min later, the left renal vein was punctured close to the hilus and 1.5 ml renal venous blood was collected at a rate of 0.1 ml/min. A second arterial blood sample (data not shown) was taken after further 15 min. Measurements were done as detailed in (6,11-13). When GSH was infused, steady state arterial plasma concentrations ranged from 0.27 to 0.69 mM. Data are means ±S.E.M.(n).

ilar to the observations (20) with S-methyl GSH and bromsulfalein-GSH, but also by an additional uptake mechanism. The capacity of this mechanism is high as shown by the experiments in which GSH was infused into the jugular vein (Table 2). Thus, even at a mean arterial plasma glutathione concentration of 0.67 mM, 79.2% of the glutathione delivered to the kidney by renal plasma flow is taken up by the renal tissue. This finding in vivo is in good agreement with an earlier study with isolated perfused rabbit kidney (21). Clearly, the limiting parameter for renal degradation of extracellular glutathione is the renal plasma flow rate.

Since, for technical reasons, the two types of experiment on the perfused liver and on the anaesthetized rat had to be carried out with animals of significantly different body weight (160 g versus 380 g), a further detailed comparison of the data appears unwarranted. However, a rough estimate shows that if the rate of GSH release into the plasma from the liver occurs as calculated above, organs other than the kidney probably also play a role in degradation of extracellular glutathione.

Acknowledgments. The expert technical assistance of C. Lachenmaier, I. Linke and A. Marklstorfer is gratefully acknowledged. This study was supported by Deutsche Forschungsgemeinschaft (Sonderforschungsbereich 51, Medizinische Molekularbiologie und Biochemie, Grant D/8; and by Grant Ha775/6).

G.M.B. was recipient of a fellowship from the Deutscher Akademischer Austauschdienst.

References

1. Binkley, F., Nakamura, K.: J. Biol. Chem. 173, 411-421 (1948)
2. Elwyn, D.H., Parikh, H.C., Shoemaker, W.C.: Am. J. Physiol. 215, 1260-1275 (1968)
3. Wendel, A., Hahn, R., Guder, W.: Renal metabolism in relation to renal function. Schmidt, U., Dubach, U.C. (eds.), pp. 426-436. Bonn, Stuttgart, Wien: Huber 1976
4. Orlowski, M., Wilk, S.: Eur. J. Biochem. 71, 549-555 (1976)
5. Hahn, R., Wendel, A., Flohé, L.: Biochim. Biophys. Acta 539, 324-337 (1978)
6. Tietze, F.: Analyt. Biochem. 27, 502-522 (1969)
7. Bartoli, G.M., Sies, H.: FEBS Lett. 86, 89-91 (1978)
8. Sies, H.: Meth. Enzymol. 52, 48-59 (1978)
9. Sies, H., Summer, K.H.: Eur. J. Biochem. 57, 503-512 (1975)

10. Racker, E.: J. Biol. Chem. 190, 685-696 (1951)
11. Oshino, N., Chance, B.: Biochem. J. 162, 509-525 (1977)
12. Häberle, D.: Kidney Int. 7, 385-396 (1975)
13. Häberle, D., Ruhland, G., Lausser, A., Moore, L., Neiss, A.: Pflüger's Arch. 375, 131-139 (1978)
14. Sies, H., Wahlländer, A., Linke, I., Marklstorfer, A.: Hoppe-Seyler's Z. physiol. Chem. 359, 1151 (1978)
15. Sies, H., Wahlländer, A., Waydhas, C.: Properties of glutathione disulfide (GSSG) and glutathione-S-conjugate release from perfused rat liver, pp.120-126 (this volume)
16. Sies, H., Gerstenecker, C., Menzel, H., Flohé, L.: FEBS Lett. 27, 171-175 (1972)
17. Higashi, T., Tateishi, N., Naruse, A., Sakamoto, Y.: J. Biochem. 82, 117-124 (1977)
18. Waelsch, H., Rittenberg, D.: J. Biol. Chem. 144, 53-58 (1942)
19. Douglas, G.W., Mortensen, R.A.: J. Biol. Chem. 222, 581-585 (1956)
20. Wendel, A., Heinle, H., Silbernagl, S.: Hoppe-Seyler's Z. physiol. Chem. 358, 1413 (1977)
21. Fonteles, M.C., Pillion, D.J., Jeske, A.H., Leibach, F.H.: J. Surg. Res. 21, 169-174 (1976)

Unidirectional Transport of Reduced Glutathione in Rat Liver and Its Metabolization in the Extracellular Space

R. HAHN and W. OBERRAUCH

Summary

Unidirectional transport for reduced glutathione (GSH) in rat liver has been found: no uptake by perfused rat liver was observed, but GSH efflux occurred at a rate of 8.6 nmol/min/g. Thus the fate of extracellular GSH was investigated.

When radioactive γ-glutamyl tripeptides were injected intravenously into rats, they were rapidly removed from the blood. Instant hydrolysis was observed. Undegraded tripeptides were detected in the blood exclusively. Within the tissues glycine accounted for all the radioactivity which accumulated first in the kidney. This was shown by autoradiographic studies and chemical analysis of the tissues. The rate of hydrolysis was drastically reduced after clamping the kidney vessels. Experiments with isolated kidney tubules revealed that γ-glutamyl transpeptidase can be considered responsible for the degrading activity.

From these observations it is concluded that extracellular hydrolysis on the luminal surface of the kidney brush border membrane and reuptake of the constituent amino acids has to be assumed for the fate of GSH released from the cell.

A. Introduction

Glutathione is released from the cell on exposure to xenobiotics (1) or "oxidative stress" (2-7). Efflux of oxidized glutathione (GSSG) has been proposed to contribute to the turnover of glutathione in erythrocytes and the eye lens (2,3,6) but experiments with erythrocyte ghosts revealed that GSSG transport does not account for the turnover (8). Recent findings however, provided evidence that not only GSSG, but also reduced glutathione is released from the cell (10). The efflux rate from perfused rat liver was found to be compatible with the turnover rate in this tissue (10). In spite of the GSH release into the extracellular space only minute amounts of the tripeptide are normally found extracellularly (11,12).

The aim of the present investigation was to investigate the efflux of reduced glutathione from liver and to elucidate the fate of extracellular GSH. For this purpose reuptake of the tripeptide by isolated rat liver was examined. Distribution and metabolization of intravenously administered radioactive tripeptides were investigated by autoradiographic studies, chemical analysis of different tissues and degradation experiments with isolated renal tubules (9,13,16).

B. Materials and Methods

I. Materials

[U-^{14}C] glycine (15 and 114 Ci/mol) and [U-^{14}C] sucrose (302 Ci/mol) were purchased from Amersham Buchler, Braunschweig. [U-^{14}C] glycine labeled GSH, GSSG and ophthalmic acid were prepared enzymatically as described earlier (13).

II. Methods

Investigation of the Efflux from Perfused Rat Liver. Liver perfusion was performed as described in (5) with Hanks medium. The GSH efflux from liver was estimated in an open perfusion system. The effluate was assayed for lactate dehydrogenase (LDH) and for total glutathione and GSSG according to (14) and (12) respectively. GSH was estimated from the difference of total and oxidized glutathione (12). Since the GSH assay is influenced by dithiothreitol (DTT), the dithiol had to be determined for correction. DTT was estimated with 5,5'-dithiobis-(2-nitrobenzoic acid) (DTNB) based on an assay described in (15). In the presence of arsenite, DTT reacts only slowly with DTNB. The time-dependent increase of the absorption was linearly related to the concentration of DTT, ranging from 0.05 mM to 0.5 mM.

Permeability Studies with Perfused Rat Liver. These studies were performed in a circulating system as described elsewhere (16). The medium containing the radioactive compound was circulated in a closed "loop" system excluding the liver (first and second circulation, C_1, C_2) in order to detect unspecific losses of radioactivity. During the "closed perfusion" (CP) the liver was included into the circulation. Thereafter the extracellular space was flushed with radioactive buffer ("open washing", OW) in an open system, which was then closed in order to detect any redistribution of radioactivity from the liver to the buffer ("closed washing", CW). The perfusion experiments with GSH were performed in the presence of 2 mM DTT.

GSH Generation from Protein-GSH Mixed Disulfides by DTT. 5 ml liver homogenate were dialyzed twice against 10 l Hanks medium for 20 h to remove acid-soluble GSH. Samples of the homogenate were rapidly brought to 0.5 mM DTT. After 5 min at 37°C and deproteinization with metaphosphoric acid, the supernatant was assayed for GSH.

Other Techniques. Autoradiographic studies, determination of intravenously injected, labeled compounds and in vitro studies with isolated renal tubules were performed as reported earlier (16).

C. Results

Release of Reduced Glutathione from Liver. When isolated rat liver was perfused with Hanks medium (perfusate flow 2.5 ml/g liver) a high GSH efflux rate was observed in the beginning of the perfusion (Fig. 1). After 20 min a constant rate of 9.4 (±1.66, n=7) nmol/min/g liver was obtained. GSSG efflux did not exceed 0.8 nmol GSH equivalents/min/g. From these data a steady-state molarity of 3.44 μM reduced glutathione in the effluate can be calculated. Since the profile of LDH release was similar to that of GSH, the initial high efflux rate can be attri-

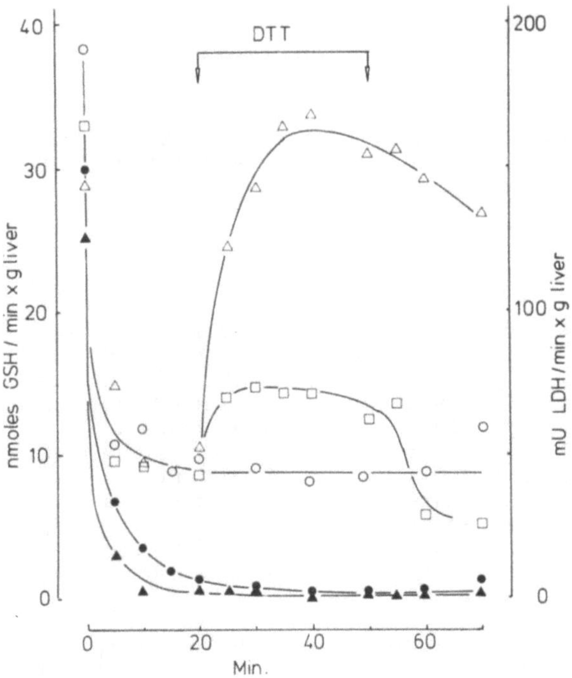

Fig. 1. GSH efflux from isolated rat liver perfused with Hanks medium without DTT (O-O) and with 0.5 mM (□-□) and 2 mM DTT (Δ-Δ). For comparison LDH efflux is shown: ●-●, perfusion without DTT; ▲-▲, with 2 mM DTT. Perfusate flow, 2.5 ml/min/g liver

buted to cell disruption during the isolation of the liver. After adaptation to the perfusion conditions, however, only 3 mU LDH/min/g were released.

Stimulation of GSH Efflux by DTT. During the steady-state release of GSH, when DTT was added to the perfusion medium a rapidly occurring extra release of up to 15 and 33 nmol GSH/min/g was observed with 0.5 and 2 mM DTT respectively, while LDH efflux remained constant. Thus the DTT-dependent efflux obviously is not connected with cell disintegration. This was further confirmed by the reversibility of the DTT-dependent GSH release (Fig. 2). The identity of the released thiol with GSH was given by (1) the specificity of the assay and (2) the finding that the blank value was obtained after preincubation with γ-glutamyl transpeptidase.

GSH Formation from Rat Liver Homogenate on Incubation with DTT. The efflux of 970 nmol GSH on perfusion with 2 mM DTT (estimated by integration of the upper curve in Fig. 1) is incompatible with an exclusive reduction of GSSG (approximately 100 nmol/g liver (12)) and subsequent transport of the corresponding GSH equivalents. Thus, protein-GSH mixed disulfides were considered as an alternative pool of oxidized glutathione.

Incubation of extensively dialyzed liver homogenate with 0.5 mM DTT at 37°C resulted in an instant GSH liberation: 186 nmol GSH were formed per gram liver within 5 min.

Permeability Studies with Perfused Rat Liver. In order to investigate the fate of glutathione released into the extracellular space, radioactive GSH was injected intravenously into rats. As a result rapid disappearance of the label from the blood stream was observed (Fig. 3).

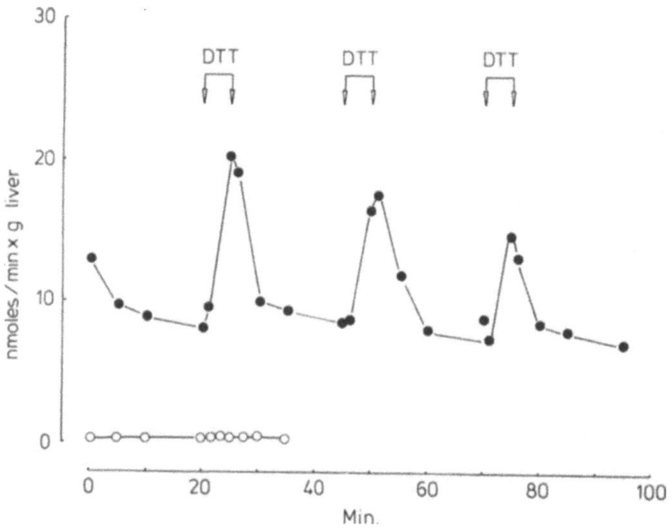

Fig. 2. Reversibility of DTT-dependent glutathione release. The liver was perfused with 0.5 mM DTT at the times indicated. ●-●, GSH; O-O, GSSG. Perfusate flow 2.5 ml/min/g liver. Each point is the mean value of two separate experiments

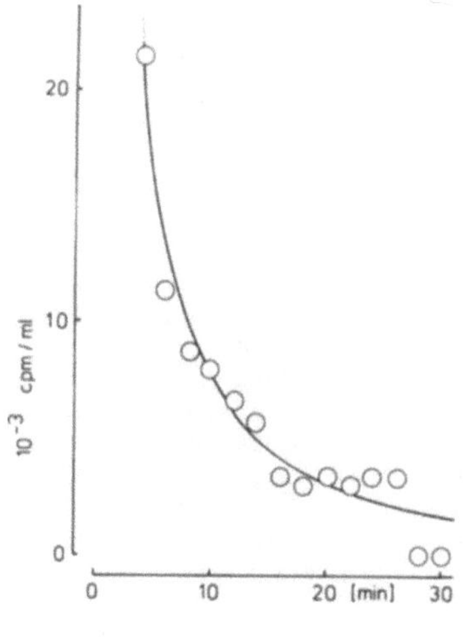

Fig. 3. Disappearance of radioactivity from the blood after intravenous administration of 1.12 µCi [*glycine*-U-^{14}C] glutathione (specific radioactivity: 0.112 Ci/mol) into a rat (16)

Thus, perfusion studies with isolated rat liver were performed in order to examine a possible reuptake of GSH.

The permeability properties of GSH were compared with those of impermeable sucrose (17) and permeable amino acids such as glycine. As shown in Figure 4 A, no significant decrease of radioactivity was observed when the liver was perfused with [U-^{14}C] sucrose (0.066 µM) in the "closed perfusion" period (CP).

The permeability of [U-^{14}C] glycine (0.22 µM) was demonstrated by the rapid removal of radioactivity from the perfusion buffer ((CP) in Fig. 4 B). In contrast to the experiments with sucrose, the radioactivity

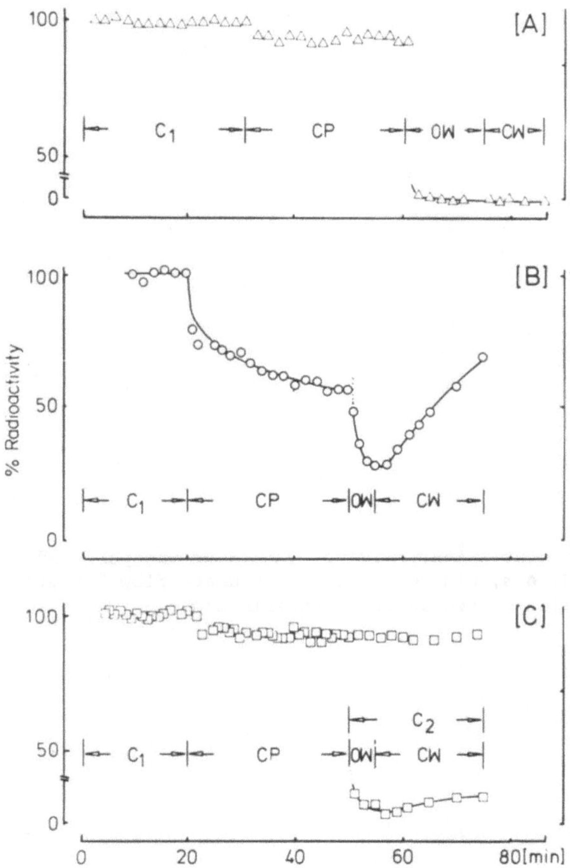

Fig. 4A-C. Hemoglobin-free perfusion of rat liver with 2.5 µCi ^{14}C-labeled sucrose, glycine and GSH. (A) [U-^{14}C] sucrose, Δ-Δ, 0.066 µM, liver weight, 10 g. (B) [U-^{14}C] glycine, O-O, 0.22 µM, liver weight, 8.8 g. (C) [*glycine*-U-^{14}C] GSH, □-□, 1.67 µM, liver wt., 10.5 g. The content of the labeled compounds in the perfusion buffer is given in % radioactivity. The radioactivity at the end of the first circulation period (C_1) was taken as 100%. Left-hand scale, first and second circulation period (C_1,C_2) and closed perfusion with 100 ml buffer originally containing the labeled compounds. Right-hand scale, open and closed washing (OW,CW) with buffer originally free of radioactivity. Oxidation of GSH was prevented by 2 mM DTT. For abbreviations see Methods. [Modified from (16)]

of the washing buffer did not reach zero within 5 min (OW) and even increased when the system was closed (CW), indicating a redistribution of labeled compounds from the liver to the perfusate. At the end of the experiment, 28% of the radioactivity remained in the liver. Very similar results were found with cysteine and glutamate (16).

The behavior of GSH (1.67 µM) is illustrated in Figure 4 C. Identical patterns were obtained with ophthalmic acid and GSSG (16). Comparison of these results with those obtained with sucrose and amino acids demonstrate that a significant uptake of γ-glutamyl tripeptides by the liver can be excluded. This assumption is supported by the finding that no loss of radioactivity occurred, as was observed with sucrose.

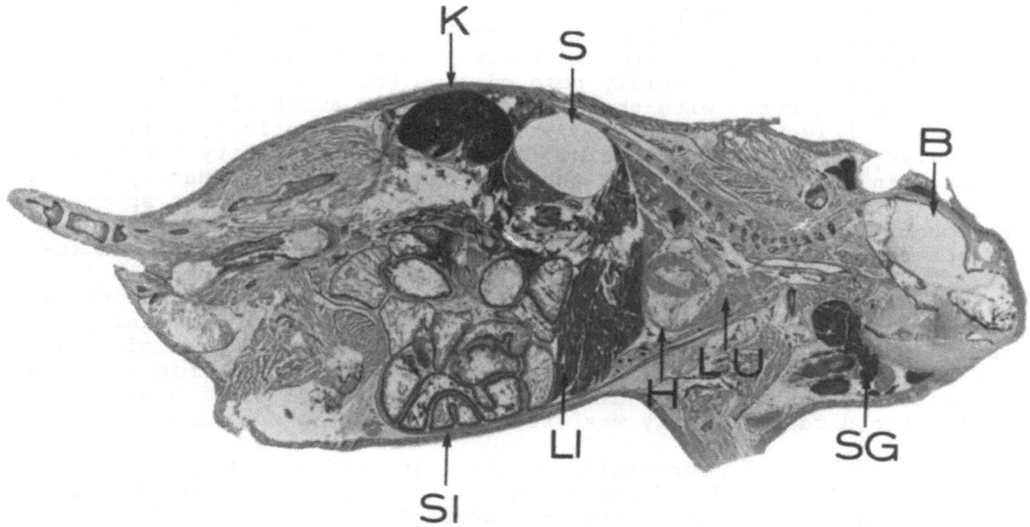

Fig. 5. Autoradiography of a mouse (body wt., 20 g) after injection of 5 µCi [*glycine*-U-^{14}C] GSSG. Following killing the animal was immediately frozen in liquid nitrogen. Freeze-cuts 15 µm thick were exposed to an X-ray film for 6 days at -20°C. B, brain; H, heart;, K, kidney; LI, liver; LU, lung; SG, salivary glands; SI, small intestine; S, stomach (16)

Table 1. Distribution of radioactivity in rat liver and kidney after intravenous injection of [^{14}C]-labeled γ-glutamyl tripeptides (16)

	Ophthalmic acid						GSH				GSSG		Glycine	
	Time of tissue removal following administration													
	3 min		3 min		30 min		5 min		30 min		5 min		30 min	
	A	B	A	B	A	B	A	B	A	B	A	B	A	B
Liver	1.00	3.31	1.00	6.52	1.00	5.70	1.00	3.82	1.00	6.93	1.00	2.14	1.00	10.00
Kidney	5.60	3.92	1.57	1.95	1.68	1.90	9.25	6.06	2.68	1.86	12.60	5.61	0.55	2.00

3, 6 and 30 min after injection of 5.3 µCi radioactive ophthalmic acid (114 mCi/nmol) into the femoral vein of rats weighing 90-120 g the liver and kidney were removed, immediately frozen in liquid nitrogen and lyophilized. An aliquot of each tissue was analyzed for radioactivity. The experiments with GSH were done with 1.12 and 2.24 µCi (0.112 and 0.224 mCi/nmol respectively) and the tissues removed after 5 and 30 min. In the experiments with GSSG 3.7 µCi (0.37 mCi/nmol) were injected and the tissues removed after 30 min. 2.5 µCi [U-^{14}C] glycine (0.25 mCi/nmol) were administered as control and the tissues removed after 30 min.
In the A columns of the table the specific radioactivities (nCi/g wet wt.) of the kidney is given in comparison to that of the liver (= 1.00). The B columns give the total radioactivity of the organs as percent of the radioactivity applied.

Autoradiographic Studies. When 3.3 µmol [*glycine*-U-^{14}C] GSSG (Fig. 5) and GSH were intravenously injected into a mouse, the kidney was the tissue showing the highest radioactivity after 10 min. In addition the salivary glands, liver and epithelium of the small intestine were radioactively labeled.

Degradation of γ-Glutamyl Tripeptides in the Rat in vivo. When [U-^{14}C] glycine labeled ophthalmic acid (50 nmol), GSH (10 μmol) and GSSG (10 μmol) were administered into the femoral vein of rats weighing 90-120 g, the radioactivity was rapidly removed from the blood (Fig. 3). The label was not excreted with the urine but spread among all tissues investigated [liver, kidney, lung, heart, spleen, testicle, epididymis, vesicular gland, small intestine (16)]. The radioactivity first accumulated in the kidney but was thereafter distributed primarily to the liver (Table 1, columns B for ophthalmic acid and GSH). The specific radioactivity of kidney was by far the highest (Table 1, columns A). In contrast, the liver was the organ containing the highest radioactivity, when [U-^{14}C] glycine was injected as control. Surprisingly only radioactive glycine was detectable in the tissues and radioactive tripeptides were found exclusively in the blood. Of the remaining radioactivity, however, only 14% was attributable to ophthalmic acid and 37% to GSH after 3 and 5 min respectively. No ophthalmic acid was detected after 6 min. GSSG disappeared more slowly: after 5 min 70% of the blood radioactivity was still detectable as the disulfide.

Consequently, rapid degradation of the tripeptides must have occurred. Since ophthalmic acid (17 μM) was not significantly metabolized on incubation with blood for 30 min, the degradation must have occurred in some tissue other than blood. When kidney arteries were clamped, the metabolization of ophthalmic acid was delayed: still 50% of the blood radioactivity was attributable to the peptide after 5 min. Thus the kidney seemed to play an important role in the degradation of extracellular γ-glutamyl tripeptides.

The degradation rate in vivo can be evaluated from Figure 3: 37% of the radioactivity represent undegraded tripeptide in the body 5 min after injection of 1.12 μCi GSH (0.112 Ci/mol). It can be calculated that 95% of the dose administered was degraded within 5 min.

Degradation of γ-Glutamyl Tripeptides by Isolated Kidney Tubules (16). In order to identify the degrading activity of the kidney, in vitro experiments with isolated renal tubules were performed. The rates of GSH decomposition were directly related to protein concentration, the specific rate was 18 nmol/min/mg protein. The degrading activity was exclusively located in the 100,000 g sediment and completely inhibited by 20 mM serine-borate, a specific inhibitor of γ-glutamyl transpeptidase (18). From these findings (16), γ-glutamyl transpeptidase was considered as the enzyme responsible for the metabolization of extracellular γ-glutamyl tripeptides. Further evidence for this assumption was provided by the formation of γ-glutamyl [U-^{14}C] glycine from radioactive glycine and non-radioactive GSH. Moreover, cystinyl-bis-[U-^{14}C] glycine was formed from [*glycine*-U-^{14}C] GSSG in addition to [U-^{14}C] glycine.

D. Conclusions

Reduced Glutathione Efflux from Perfused Rat Liver. The GSH efflux rate of 8.6 nmol/min/g is slightly below the value reported in (10). The cause for enhanced GSH release from rat liver by DTT is unknown. Although an unspecific effect on the permeability barrier cannot be excluded, a DTT-induced leakage of the cell is unlikely for the following reasons: (1) no concomitant appearance of LDH was observed, (2) the DTT-dependent efflux was reversible and repeatable and (3) no radioactive GSH was incorporated from the perfusion medium into the intracellular GSH pool in the presence of 2 mM DTT.

The unidirectional transport of GSH may be considered as a regulatory event upon an increase of the 2 GSH/GSSG ratio. Reduction of only GSSG would not generate sufficient GSH to explain the GSH efflux on DTT infusion. Protein-GSH mixed disulfides, however, have long been suggested as GSH pool (19). Thus, reduction of mixed disulfides through DTT may have led to an increased 2 GSH/GSSG ratio and possibly to an extra release of GSH. Clearly, the question whether this can take place under physiological conditions remains to be answered.

The Fate of Intravenously Injected γ-Glutamyl Tripeptides. The results reported here demonstrate that liver is essentially impermeable to exogenous GSH, GSSG and ophthalmic acid: (1) no uptake in perfusion experiments was obtained and (2) no labeled peptides were found in liver shortly after injection. Consequently, the liver does not seem to be of importance in the rapid removal of extracellular GSH as was observed in vivo. More likely, degradation and subsequent resorption of the constituent amino acids plays an important role for the disappearance of GSH released into the extracellular space. This assumption is evidenced by the finding that rapid degradation of γ-glutamyl tripeptides occurred after intravenous injection and glycine only was attributable to the radioactivity in the tissues.

The kidney appears to be of great significance in the decomposition of GSH since it was the tissue with the highest radioactivity immediately after administration of labeled tripeptides, and in particular the degradation was markedly delayed when the renal arteries were clamped. The rapid degradation of intravenously administered GSH is incompatible with the maintenance of a steady-state concentration of approximately 2.6 μmol/g (20,21) and a half-life of 29 min (21) in the kidney and thus with the common assumption that GSH is metabolized exclusively intracellularly (22,23). On the contrary, an extracellular metabolization has to be considered. Evidence is provided by the observation that undegraded labeled γ-glutamyl tripeptides were not found in the kidney, not even immediately after administration, but only in the blood. Since the degrading activity of the kidney can be attributed to γ-glutamyl transpeptidase, which is localized on the kidney brush border membrane (24), it can be concluded that GSH degradation has occurred on the luminal surface of the renal tubule and γ-glutamyl transpeptidase acts as a glutathionase. The majority of γ-glutamyl transpeptidase thus does not seem to be in contact with intracellular GSH.

Finally, release from the cell, extracellular hydrolysis, and reuptake of the constituent amino acids with subsequent resynthesis of GSH may be considered as interorgan turnover. Whether this is of physiological significance remains to be elucidated. Evidence is provided, however, by the finding of elevated levels of GSH in blood and urine of a patient deficient in γ-glutamyl transpeptidase (11).

References

1. Boyland, E., Chasseaud, L.F.: The role of glutathione and glutathione S-transferase in mercapturic acid biosynthesis. Adv. Enzym. 32, 173-219 (1975)
2. Srivastava, S.K., Beutler, E.: Permeability of normal and glucose-6-phosphate dehydrogenase deficient erythrocytes to glutathione. Biochem. J. 112, 421-425 (1969)
3. Srivastava, S.K., Beutler, E.: Cataract produced by tyrosinase and tyrosine systems in rabbit lens in vitro. Biochem. J. 112, 421-425 (1969)

4. Sies, H., Gerstenecker, C., Menzel, H., Flohé, L.: Oxidation in the NADP system and release of GSSG from hemoglobin-free perfused rat liver during peroxidatic oxidation of glutathione by hydroperoxides. FEBS Lett. 27, 171-175 (1972)
5. Sies, H., Summer, K.H.: Hydroperoxidase-metabolizing systems in rat liver. Eur. J. Biochem. 57, 503-512 (1975)
6. Smith, J.E.: Relationship of in vivo erythrocyte glutathione flux to the oxidized glutathione transport system. J. Lab. Clin. Med. 83, 444-450 (1974)
7. Oshino, N., Chance, B.: Properties of glutathione release observed during reduction of organic hydroperoxide, demethylation of aminopyrine and oxidation of some substances in perfused rat liver, and their implications for the physiological function of catalase. Biochem. J. 162, 509-525 (1977)
8. Prchal, J., Srivastava, S.K., Beutler, E.: Active transport of GSSG from reconstituted erythrocyte ghosts. Blood 46, 111-117 (1975)
9. Hahn, R.: Utilisation von extrazellulärem Glutathion in der Ratte. Ph.D. Thesis, Univ. Tübingen (1975)
10. Bartoli, G.M., Sies, H.: Reduced and oxidized glutathione efflux from liver. FEBS Lett. 86, 89-91 (1978)
11. Schulman, J.D., Goodman, S.I., Mace, J.W., Patrick, A.D., Tietze, F., Butler, E.J.: Glutathionuria: inborn error of metabolism due to tissue deficiency of gamma-glutamyl transpeptidase. Biochem. Biophys. Res. Commun. 65, 68-74 (1975)
12. Tietze, F.: Enzymatic method for quantitative determination of nanogram amounts of total and oxidized glutathione: Applications to mammalian blood and other tissues. Anal. Biochem. 27, 502-522 (1969)
13. Wendel, A., Hahn, R., Guder, W.G.: On the role of γ-glutamyltransferase in renal tubular amino acid reabsorption. In: Renal metabolism in relation to renal function. Schmidt, U., Dubach, U.C. (eds.), pp. 426-436. Bern-Stuttgart-Vienna: Hans Huber Publishers 1976
14. Brehe, J.E., Burch, H.B.: Enzymatic assay for glutathione. Anal. Biochem. 74, 189-197 (1976)
15. Zahler, W.L., Cleland, W.W.: A specific and sensitive assay for disulfides. J. Biol. Chem. 243, 716-719 (1968)
16. Hahn, R., Wendel, A., Flohé, L.: The fate of extracellular glutathione in the rat. Biochim. Biophys. Acta 539, 324-337 (1978)
17. Baur, H., Heldt, H.W.: Glucose transport by isolated hepatocytes. In: Use of isolated liver cells and kidney tubules in metabolic studies. Tager, J.M., Söling, H.D., Williamson, J.R. (eds.), pp. 357-563. Amsterdam: North Holland Publishing Co. 1976
18. Revel, J.P., Ball, E.G.: The reaction of glutathione with amino acids and related compounds as catalyzed by γ-glutamyl transpeptidase. J. Biol. Chem. 234, 577-582 (1959)
19. Modig, H.: Cellular mixed disulphides between thiols and proteins and their possible implication for radiation protection. Biochem. Pharmacol. 17, 177-186 (1968)
20. Palekar, A.G., Tate, S.S., Meister, A.: Decrease in glutathione levels of kidney and liver after injection of methionine sulfoximine into rats. Biochem. Biophys. Res. Commun. 62, 651-657 (1975)
21. Sekura, R., Meister, A.: Glutathione turnover in the kidney; considerations relating to the γ-glutamyl cycle and the transport of amino acids. Proc. Natl. Acad. Sci. U.S.A. 71, 2969-2972 (1974)
22. Meister, A.: On the enzymology of amino acid transport. Science 180, 33-39 (1973)
23. Meister, A., Tate, S.S., Ross, L.L.: Membrane bound γ-glutamyl transpeptidase. In: The enzymes of biological membranes. Martonosi, A. (ed.), Vol.III, pp. 315-347. London: Wiley 1976
24. Glossmann, H., Neville, D.M.: γ-Glutamyltransferase in kidney brush border membranes. FEBS Lett. 19, 340-344 (1971)

Chapter II
The Role of γ-Glutamyl-Transferase in Glutathione Turnover

Editor's Note: In this book, the terms γ-glutamyl-transferase (E.C.2.3.2.2) and γ-glutamyl-transpeptidase are used interchangeably.

Current Status of the γ-Glutamyl Cycle

A. MEISTER

Summary

According to the γ-glutamyl cycle, the first step in glutathione breakdown is catalyzed by γ-glutamyl transpeptidase, which transfers the γ-glutamyl moiety of glutathione to an acceptor amino acid to form a γ-glutamyl amino acid. γ-Glutamyl amino acids may be hydrolyzed, participate themselves in transpeptidation, or be converted by γ-glutamyl cyclotransferase to 5-oxoproline and the corresponding amino acids. 5-Oxoproline is converted to glutamate in a reaction coupled with ATP cleavage to ADP and inorganic phosphate. Cysteinyl-glycine, formed in the transpeptidation reaction, is split to cysteine and glycine. Glutamate, cysteine, and glycine are converted to glutathione in two successive steps catalyzed, respectively, by γ-glutamylcysteine and glutathione synthetases. Metabolite labeling studies and experiments with specific inhibitors that block the cycle at various steps showed that the reactions of the cycle take place in vivo. Studies with enzyme inhibitors and other data, including information obtained from observations on patients with inborn blocks of the cycle, indicate that the γ-glutamyl cycle accounts for the turnover of intracellular glutathione and functions as one of the systems that mediates translocation of amino acids. Recent in vivo studies on the effect of amino acid administration on glutathione and 5-oxoproline levels, and on the effects of specific inhibition of glutathione synthesis and γ-glutamyl transpeptidase indicate that there is a physiologically significant connection between the metabolism of glutathione and amino acid transport.

A. Introduction

The γ-glutamyl cycle, the metabolic pathway that accounts for the synthesis and degradation of glutathione in many mammalian tissues, was elucidated by investigations in this laboratory on the enzymes that catalyze glutathione synthesis and by studies which showed that 5-oxoproline is a metabolite, formed from the γ-glutamyl moiety of glutathione, and converted to glutamate in an ATP-requiring reaction (24, 28). These studies, the details of which have been recently reviewed (28), also showed that 5-oxoproline is a normal constituent of tissues and body fluids. The first step in the degradation of glutathione is catalyzed by γ-glutamyl transpeptidase, an enzyme that transfers the γ-glutamyl moiety of glutathione to an amino acid acceptor to form a γ-glutamyl amino acid. γ-Glutamyl amino acids may be hydrolyzed, undergo additional transpeptidation reactions, or serve as substrates of γ-glutamyl cyclotransferase, which converts them to 5-oxoproline and the corresponding amino acids. 5-Oxoproline is converted to glutamate in a reaction coupled with the cleavage of ATP to ADP and inorganic phosphate. Cysteinylglycine, formed in the transpeptidation reaction of glutathione, is cleaved by dipeptidase to glycine and cysteine. Glutathione is synthesized in two steps from its three constituent amino acids; these reactions are catalyzed, respectively, the γ-glutamylcysteine synthetase and glutathione synthetase. Prior to our work on the γ-glutamyl cycle, the significance of γ-glutamyl cyclotransferase was not understood and 5-oxoproline was generally regarded as an artifact. The finding that 5-oxoproline is rapidly metabolized in vivo, and the subsequent discovery of the enzyme 5-oxoprolinase thus provided a metabolic raison d'être for γ-glutamyl cyclotransferase. These developments indicated that there is a metabolic connection between the synthesis and the degradation of glutathione, and thus made it possible to visualize the γ-glutamyl cycle.

There is good evidence that the reactions of the γ-glutamyl cycle occur in vivo. The enzymes and substrates of the cycle are widely distributed. The reactions of the cycle have been demonstrated in vivo by labeling studies and by use of specific enzyme inhibitors. The turnover of glutathione is much higher in kidney than it is in liver (46), a finding consistent with the much higher activities of the cycle enzymes in kidney than liver. Administered 5-oxoproline and glutamate are rapidly incorporated (at about the same rate) into glutathione. Administration of labeled glutamate is followed by prompt appearance of the label in kidney and liver 5-oxoproline (48). Animals treated with a competitive inhibitor of 5-oxoprolinase accumulate 5-oxoproline in their tissues, and administration of labeled 5-oxoproline to animals thus treated reveals a marked decrease in the usually rapid conversion of 5-oxo[^{14}C] proline to $^{14}CO_2$ (64). Animals given the inhibitor together with amino acids show increased 5-oxoproline accumulation, which can be explained by the operation of the γ-glutamyl cycle, since administration of amino acids would be expected to increase transpeptidation, and thus to increase 5-oxoproline formation. Further evidence that the reactions of the cycle occur in vivo has come from studies on patients with 5-oxoprolinuria. In this disease, glutathione synthetase is markedly decreased leading to a deficiency of glutathione (66). Since glutathione serves normally as a feedback inhibitor of γ-glutamylcysteine synthetase (42), there is an increase in the formation of γ-glutamylcysteine, which in turn, is rapidly converted to 5-oxoproline and cysteine. The formation of 5-oxoproline exceeds the capacity of 5-oxoprolinase, and 5-oxoproline therefore accumulates and is excreted in large amounts in the urine. Patients with 5-oxoprolinuria thus have a modified γ-glutamyl cycle involving the actions of four enzymes (25,26). More recent studies provide additional evidence that the cycle functions in vivo (Chap.C. VIII).

B. Functions of the γ-Glutamyl Cycle

In proposing that the γ-glutamyl cycle functions as one of the systems that mediates the transport of amino acids across cell membranes, we considered several observations (24,28,34). Thus, the breakdown of glutathione catalyzed by γ-glutamyl transpeptidase is greatly stimulated by certain amino acids, which are not metabolized in the cycle but are released unchanged. γ-Glutamyl transpeptidase is bound to the membranes of certain epithelial cells at sites known to be extensively involved in transport (e.g., jejunal villi, proximal renal tubules, choroid plexus, ciliary body, visual receptor cells, retinal epithelium, cerebral astrocytes or their capillaries, lymphoid cells (29,32)), whereas the other enzymes of the cycle are present in the cytosol. Earlier kinetic studies on amino acid transport led to the proposal that this process involves several steps, e.g., binding of amino acids to a cell membrane site, carrier-mediated translocation, intracellular release of amino acid from carrier, and energy-dependent reactivation of carrier (17,39). The γ-glutamyl cycle thus has features previously postulated to be involved in amino acid transport. The membrane-bound transpeptidase might mediate binding of amino acids and their translocation, by interacting with extracellular amino acid and intracellular glutathione (or possibly another γ-glutamyl donor derived from glutathione) to yield a γ-glutamyl amino acid. According to this interpretation, the γ-glutamyl moiety is the carrier and the γ-glutamyl amino acid is translocated into the cell. Within the cell, the amino acid is released from its γ-glutamyl carrier by the action of γ-glutamyl cyclotransferase. The energy-requiring steps of the cycle include those involved in the resynthesis of glutathione and the conversion of 5-oxoproline to glutamate.

It should be emphasized that we have not proposed that the γ-glutamyl cycle mediates transport of all amino acids, nor have we suggested that it is the only amino acid transport mechanism. Evidence derived from many approaches indicates that there are a multiplicity of amino acid transport mechanisms; we have proposed that the γ-glutamyl cycle functions as one of these. It may be noted that the γ-glutamyl cycle is thus far the only amino acid translocation mechanism in which specific enzymes have been implicated. Although there are undoubtedly other transport mechanisms for amino acids, the information available about these is indirect and based largely on measurements of uptake rates and the observation of competitive effects. In no instance has information been obtained about the chemical phenomena involved in such "systems", which must therefore be regarded as hypothetical. It is clear, however, that energy is needed for active transport and that the postulated systems exhibit "overlapping" specificity.

The amino acid acceptor specificity of γ-glutamyl transpeptidase is such as to suggest that certain neutral amino acids (including glutamine and cystine) are preferentially handled by the γ-glutamyl cycle. Glutamine may play a special role; the high concentration of glutamine in body fluids and tissues suggests that this amino acid is readily converted to γ-glutamylglutamine. Indeed there is evidence for the presence of γ-glutamylglutamine in many mammalian tissues (19). Glutamine is an excellent acceptor amino acid substrate of γ-glutamyl transpeptidase, and γ-glutamylglutamine is the best γ-glutamyl amino acid substrate of γ-glutamyl cyclotransferase. Glutamine may also serve as a γ-glutamyl donor for γ-glutamyl transpeptidase. Although the enzyme exhibits relatively low affinity in vitro for glutamine as a γ-glutamyl donor as compared to glutathione, glutamine might serve as a γ-glutamyl donor under certain physiological conditions, or if the affinity of the enzyme for glutamine were increased by interaction with a compound that alters its conformation.

Since cystine is an excellent acceptor substrate of γ-glutamyl transpeptidase, we have considered the possibility that the transport of this amino acid may be effected by the cycle; its transport may be integrated with other aspects of the metabolism of this amino acid (58). Various modifications of the cycle have been considered (28); a γ-glutamyl cycle involving cleavage of only 2 ATP molecules might take place if the formation of 5-oxoproline were by-passed by hydrolysis. A cycle involving cleavage of only one molecule of ATP is also conceivable provided there are succesive transpeptidation reactions, one of which involves cysteine. We have also considered a model for exchange diffusion in which γ-glutamyl amino acid complexes are involved. Amino acids such as aspartate and proline, which are poor substrates for transpeptidase, are most likely transported largely by other mechanisms.

The γ-glutamyl cycle may also have other functions (28). For example, γ-glutamyl transpeptidase probably functions in the removal of the γ-glutamyl group of S-substituted glutathione derivatives which are formed in the pathway leading to mercapturic acids (see Chap.C.VIII). Other types of membrane phenomena, including protection mechanisms and neurotransmission, may also involve the γ-glutamyl cycle. The cycle functions in the conversion of free cysteine to an apparently less metabolically active tripeptide form and also in the release of cysteine from glutathione. Thus, the intracellular level of cysteine is extremely low, and a rapid release of cysteine from glutathione (a process that would be stimulated by other amino acids) might be necessary, under certain circumstances, for protein synthesis.

C. Recent Studies on the γ-Glutamyl Cycle

We have pursued studies on the cycle by means of several related approaches. The enzymes have been purified and further characterized and their mechanisms of action have been elucidated. The specificities of the enzymes have been examined in detail in an effort to design specific inhibitors that are active in vivo. Substrate analogs have been sought that can function in some, but not all of the steps of the cycle. In vivo studies have been performed in which the effects of amino acid administration on the tissue levels of glutathione and 5-oxoproline were determined. Some of these studies have been carried out in the presence of specific enzyme inhibitors.

I. γ-Glutamyl Transpeptidase

Highly purified preparations of this enzyme have been obtained from kidney and other sources (28). Thus, a preparation exhibiting a molecular weight of about 70,000 and containing about 19% carbohydrate was obtained in this laboratory from rat kidney (53). Treatment of the enzyme with sodium dedocyl sulfate in the presence of 2-mercaptoethanol gave two unequal glycopeptide subunits of molecular weight 23,000 and 46,000 (54). A similar preparation isolated from human kidney was found to exhibit a molecular weight of about 84,000; this enzyme is composed of two nonidentical glycopeptides exhibiting molecular weights of 22,000 and 62,000 (57). Both the human and rat kidney enzymes are rapidly inactivated by the γ-glutamyl analogs 6-diazo-5-oxo-L-norleucine and L-azaserine (55,57) and such inactivation is prevented by γ-glutamyl substrates. 6-Diazo-5-oxo-norleucine was shown to react specifically, covalently, and stoichiometrically at the γ-glutamyl site of the enzyme which was found to be associated with the 22,000 molecular weight subunit of both the rat (55) and human (57) kidney enzymes. These studies and kinetic investigations on the enzyme are consistent with a ping-pong mechanism involving two half-reactions (52). Since the γ-glutamyl enzyme can react with water or with another molecule of γ-glutamyl donor, the enzyme can catalyze hydrolysis, autotranspeptidation, as well as transpeptidation (59). Since D-γ-glutamyl compounds are γ-glutamyl substrates, whereas D-amino acids are not acceptor substrates, it was possible to design assay systems in which autotranspeptidation cannot occur (59). The γ-glutamyl donor site of the enzyme exhibits low stereospecificity (but high affinity), while the acceptor site exhibits absolute L-specificity. In addition to various chromogenic substrates, S-substituted glutathione derivatives are also active γ-glutamyl donors. The S-pyruvoyl and S-acetophenone derivatives of glutathione have been useful in spectrophotometric assays (52). S-acyl derivatives of glutathione are also active; the S-acyl-cysteinylglycine product formed rearranges spontaneously to the corresponding N-acyl derivative liberating a product with a free SH group which can be quantitavely determined (50).

The relative effectiveness of the transpeptidase in catalyzing transpeptidation as compared to hydrolysis of γ-glutamyl compounds can be influenced in vitro by the experimental conditions employed. Maleate is known to stimulate the hydrolysis and to decrease the transpeptidation of many γ-glutamyl compunds (51). The recent discovery (61) that the naturally occurring compound hippurate also exhibits this effect may be of physiological significance. Both hippurate and p-aminohippurate decrease transpeptidation and increase hydrolysis of many γ-glutamyl compounds by γ-glutamyl transpeptidase. Kinetic studies indicate that hippurate, maleate, and related compounds bind effectively and reversibly to the cysteinylglycine binding site of the enzyme.

γ-Glutamyl transpeptidase is inhibited by L-serine in the presence of borate (40). It is probable that en enzyme-borate-serine complex is formed that mimics the structure of the γ-glutamyl enzyme or, more likely, that of the transition state (56). The enzyme is also inhibited by a number of γ-glutamyl hydrazones of α-keto acids (52). Recent studies on a series of γ-glutamyl hydrazides indicate that this class of compound includes a number of highly active transpeptidase inhibitors (12).

The most active amino acid acceptors of the transpeptidase are neutral L-amino acids such as glutamine, cystine, methionine, asparagine, serine, and alanine. Amino acids similar in structure to cystine, such as cystathionine and α-ε-diaminopimelic acid, also exhibit activity (58). Glycylglycine is an excellent acceptor substrate (16). Dipeptides of the X-gly type are almost always substantially more active than dipeptides of the gly-X type (60,30). This suggests that aminoacyl glycines bind to the portion of the glutathione binding site of the enzyme that attaches to cysteinylglycine. The relative activity observed with peptides of the X-gly type fall in an order which is essentially parallel to that found with the corresponding free amino acids, suggesting that free amino acids bind to the site that normally binds the cysteinyl moiety of cysteinylglycine, and that the presence of a C-terminal glycine residue in the dipeptide assists substantially in binding to the enzyme.

II. γ-Glutamyl Cyclotransferase

The activity of this enzyme was first observed in 1942 by Woodward and Reinhart (69), who found that both peptide bonds of glutathione are cleaved by rat kidney extracts to yield glycine, cysteine, glutamate, and 5-oxoproline. Connell and Hanes (4) later found an activity in liver that catalyzes the conversion of γ-glutamyl amino acids to 5-oxoproline and the corresponding amino acid. As discussed above, the metabolic significance of this enzyme became apparent after it was recognized that 5-oxoproline is a normal metabolite. Highly purified preparations of γ-glutamyl cyclotransferase have been obtained from human and sheep brain (37), hog liver (1), rat liver (36), and rat kidney (49). In these studies a number of forms of the enzyme were obtained which were separable by ion exchange chromatography, isoelectric focusing, and starch gel electrophoresis. Purified preparations of the enzyme were found to be unstable and to undergo substantial changes in their physical and catalytic properties during preparation and storage. Recent studies in our laboratory on the enzyme from rat kidney (49) showed that the enzyme contains sulfhydryl groups whose modification is associated with the appearance or disappearance of the various enzyme forms. Very highly active and apparently homogeneous forms of the enzyme were obtained by chromatography on Thiol-Sepharose, and also by treatment with dithiothreitol followed by reaction with iodoacetamide. The specificity of highly purified rat kidney γ-glutamyl cyclotransferase was studied with a variety of γ-glutamyl compounds, and found to act most effectively on γ-glutamylglutamine. It is also very active toward many di-γ-glutamyl amino acids. Thus, the preferred substrates have the structure, γ-glu-γ-glu-NHR, in which the nature of the R moiety appears to have relatively little effect on activity (49). These studies suggest that the major physiological function of the enzyme is related to its activity toward γ-glutamylglutamine and possibly also toward di-γ-glutamyl amino acids.

III. 5-Oxoprolinase

This cytosolic enzyme activity has been found in a number of mammalian tissues including kidney, lung, spleen, liver, brain, ciliary body, and testes (62,63). The most active preparation of the enzyme thus far obtained from rat kidney is about 1600-fold purified. The equilibrium between 5-oxoproline and glutamate lies greatly in favor of cyclization (27,68). In the 5-oxoprolinase reaction, hydrolysis of 5-oxo-L-proline is coupled with hydrolysis of ATP and the overall reaction proceeds essentially to completion. The enzyme is competitively inhibited by L-2-imidazolidone-4-carboxylate and L-dihydroorotic acid (62,65). Although the 5-oxoprolinase reaction proceeds with almost exact stoichiometry between 5-oxoproline and ATP hydrolysis over a broad range of pH, uncoupling of the reaction occurs when 5-oxoproline or ATP is replaced by analogs. When ATP is replaced by ITP, GTP, or UTP, the hydrolysis of the nucleoside triphosphate exceeds that of the amide by 6- to 50-fold (9). In the absence of 5-oxoproline, the enzyme catalyzes a slow ATPase reaction, but it catalyzes very rapid ITPase, GTPase, and UTPase reactions. These NTPase reactions, which may be more rapid than the ATP-mediated overall coupled reaction, are inhibited by 5-oxoproline and certain 5-oxoproline analogs that bind to the enzyme.

The mechanism of the reaction has been studied with 5-oxo-L-proline containing ^{18}O in the amide carbonyl group (14). When the enzyme was incubated with such labeled 5-oxo-L-proline, the glutamate formed was found to contain all of the isotope originally present in the labeled 5-oxo-L-proline. When the enzyme was incubated with unlabeled 5-oxo-L-proline and ATP in the presence of ^{18}O-water, only one oxygen atom of the glutamate was found to be derived from ^{18}O-water. The findings are consistent with a mechanism involving phosphorylation of the amide carbonyl group of 5-oxoproline to give an enzyme-bound intermediate that can add water to form enzyme-bound γ-glutamyl phosphate, which is hydrolyzed.

IV. γ-Glutamylcysteine Synthetase

This widely distributed enzyme has been obtained in highly purified form from rat kidney (35,43,47). The most highly purified preparation was obtained by a method of purification involving chromatography on Sepharose-amino-hexyl-ATP (47). This preparation is inhibited by and binds one mol of methionine sulfoximine phosphate per mol of enzyme. The enzyme is also inhibited by L-2-amino-4-oxo-5-chloropentanoate, and such inhibition is associated with binding of a stoichiometric amount of this chloroketone inhibitor. Studies on mapping of the glutamate site of the enzyme have been carried out and several inhibitors have been obtained (10). It is notable that β-glutamate and N-methyl-L-glutamate are substrates. The enzyme is markedly inhibited by prothionine sulfoximine (see Chap.C.VII).

V. Glutathione Synthetase

Glutathione synthetase has been purified from yeast (31) and bovine (67) and human erythrocytes (23). The specific activity of the yeast enzyme is considerably higher than that of the purified erythrocyte enzymes. However, a highly purified preparation of glutathione synthetase obtained from rat kidney (33) exhibits an activity similar to that of the yeast enzyme. Recent studies on glutathione synthetase (33) show that the pattern of specificity exhibited by this enzyme differs markedly from that shown by γ-glutamylcysteine synthetase. The latter enzyme exhibits relatively strict specificity toward glutamate analogs, but uses a number of amino acids in place of cysteine. Glutathione synthetase has a very broad specificity toward glutamyl-modified analogs of γ-glutamylcysteine, but a more restricted specificity toward cysteine-modified analogs. The glutamyl site of glutathione synthetase can accept D-glutamyl, β-aminoglutaryl, or any of a number of monomethyl-substituted glutamyl residues. Recent studies show that even more drastic modification of substrate is possible; thus, N-acetyl-L-α-aminobutyrate and N-acetyl-L-cysteine also exhibit appreciable activity. In contrast, replacement of the cysteine moiety of γ-glutamylcysteine by N-methyl-α-aminobutyrate, α-methyl-α-aminobutyrate, β-aminoisobutyrate, β-amino-n-butyrate, or D-α-aminobutyrate leads to loss of activity. Although γ-glutamyl-α-aminobutyrate is as active as γ-glutamylcysteine, the corresponding glycine, alanine, and norvaline derivatives are much less active.

VI. Selective Inhibition of γ-Glutamyl Cycle Enzymes by Substrate Analogs

We have looked for specific inhibitors of the individual reactions of the cycle and also substrate analogs that would function in some but not all of the reactions. Most of these reactions involve the γ-carboxyl group of glutamate and therefore a given substrate analog might be expected to interact with more than one enzyme. However, the reaction pathways of these enzymes are quite different and are presumably associated with differences in the enzyme-bound conformations of the glutamate carbon chain at the active sites of the several enzymes. Our studies (10) have shown that suitable modification of the glutamyl moi-

	ENZYMES				
	❶	❷	❸	❹	❺
Substrate Analogs	γ-Glu-CysH Synthetase	GSH Synthetase	γ-Glu Transpeptidase	γ-Glu Cyclotransferase	5-Oxoprolinase
D-Glu	0,↓	+	+	0,↓	0
α-Methyl-Glu	(+)	+	+	0	0
N-Methyl-Glu	+	+	0	0	0
β-Glu	+	+	0	0,↓	
β-Methyl-Glu	(+),↓	+,↓	0	+	0,↓
γ-Methyl-Glu	0,↓	+	0	0	(+),↓

↓ = >50% inhibition; + = >10% as active as L-glu;
(+) = 1–10% as active as L-glu; 0 = <1% as active as L-glu

Fig. 1. Interaction of glutamate and glutamyl analogs with enzymes of the γ-glutamyl cycle [from 10] ety of the substrates can lead to the desired results, namely specific inhibitors or nonmetabolizable analogs at each of the steps of the γ-glutamyl cycle (Fig. 1). The data obtained suggest a number of fruitful experimental pathways. Thus, in one approach, β-glutamyl-L-α-aminobutyrate was found to be a highly active competitive inhibitor of γ-glutamyl cyclotransferase. In vivo studies with this compound are summarized below (Chap.VII). In addition to those summarized in Figure 1, a number of other selective inhibitors are currently available; these include methionine sulfoximine and related compounds, several inhibitors of γ-glutamyl transpeptidase (see above), and L-2-imidazolidone-4-carboxylate.

VII. Blocks of the Cycle Produced by Enzyme Deficiencies and Inhibitors

Figure 2 summarizes the effects of several inborn errors of metabolism and various inhibitors on the enzymes of the cycle. Patients with 5-oxoprolinuria have a marked generalized deficiency of glutathione synthetase, and exhibit increased formation of γ-glutamylcysteine which is converted to 5-oxoproline (25,26,42,66). This block evidently does not stop the cycle because γ-glutamylcysteine is a substrate of the transpeptidase. Patients with γ-glutamylcysteine synthetase deficiency (20,41) have a markedly reduced capacity to synthesize both glutathione and its immediate precursor. Such a defect, if complete, would be expected to stop the function of the γ-glutamyl cycle. Such patients exhibit generalized aminoaciduria, a finding consistent with the proposed role of the cycle in amino acid transport. A patient reported to have a marked deficiency of γ-glutamyl transpeptidase activity (45) had higher than normal excretions of glutamine and other amino acids

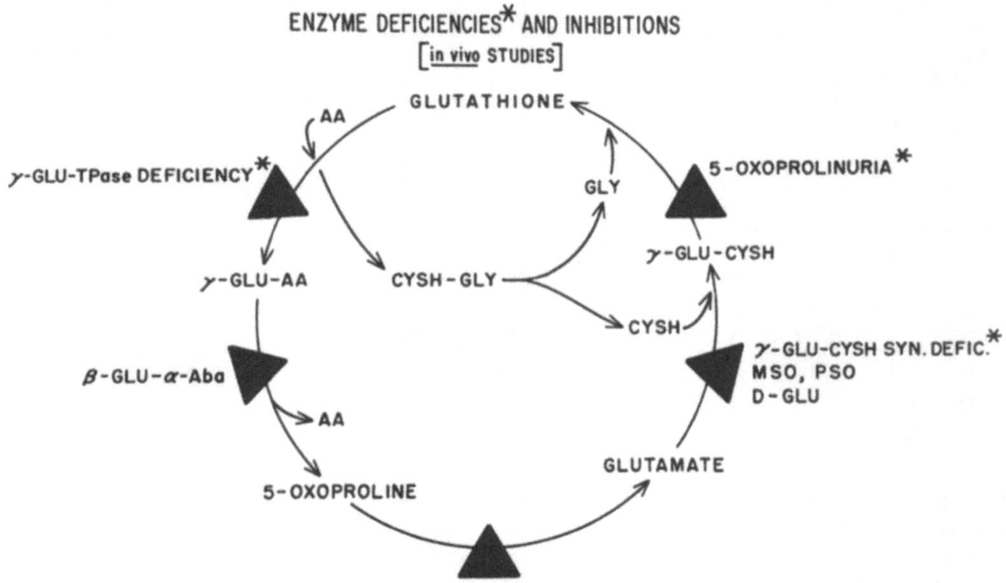

Fig. 2. Blocks of the γ-glutamyl cycle; MSO, methionine sulfoximine; PSO, prothionine sulfoximine

(28), but did not have a marked aminoaciduria. Possibly this patient, 37 years of age, has adapted to his enzyme deficiency; since there is evidence for multiplicity of amino acid transport systems it is conceivable that the patient has accomodated to the loss of one system by hypertrophy of others (28). However, it is notable that this patient exhibits substantial glutathionemia and glutathionuria; this is considered below.

Inhibition of 5-oxoprolinase by L-2-imidazolidone-4-carboxylate, a competitive inhibitor with respect to 5-oxoproline, produces 5-oxoproline accumulation in vivo (64). Similarly, administration of β-glutamyl-α-aminobutyrate to mice is followed by a significant depression of the steady-state concentration of 5-oxoproline in the kidney (3,8). Administration of β-glutamyl-α-aminobutyrate also prevents the accumulation of 5-oxoproline which is observed after administration of large amounts of amino acids. These studies, in which imidazolidone carboxylate and β-glutamyl-α-aminobutyrate were injected into mice, support the view that γ-glutamyl cyclotransferase and 5-oxoprolinase are major in vivo catalysts for the formation and utilization, respectively, of 5-oxo-L-proline.

In vivo inhibition of γ-glutamylcysteine synthetase occurs after administration of methionine sulfoximine (38). Investigations in our laboratory showed that methionine sulfoximine inhibits glutamine synthetase (5,28,44), and γ-glutamylcysteine synthetase (43) by serving as an inhibitory analog of the tetrahedral intermediates or transition states formed on these enzymes in the attack of ammonia or cysteine on enzyme-bound γ-glutamyl phosphate. Injection of methionine sulfoximine produces marked decreases in tissue levels of both glutathione and glutamine. The inhibitory effect of methionine sulfoximine on glutamine biosynthesis (and its associated convulsant activity) limits the usefulness of this reagent as an inhibitor of glutathione biosynthesis. However, recent studies in which differences between the active sites of glutamine synthetase and γ-glutamylcysteine synthetase were considered, suggested the possibility of designing and synthesizing analogs of methionine sulfoximine that would selectively inhibit each of the synthetases. These studies (7,11) led to the finding that α-ethylmethionine sulfoximine, which is a convulsant, inhibits glutamine synthetase effectively, but has no significant effect on γ-glutamylcysteine synthetase (11). Methionine sulfoximine analogs in which the S-methyl group is replaced with much bulkier moieties do not inhibit glutamine synthetase appreciably, but markedly inhibit γ-glutamylcysteine synthetase (7); such compounds are not convulsants. Thus, when prothionine sulfoximine (S-propyl homocysteine sulfoximine; PSO) is injected into mice, there is a marked decrease in the level of glutathione in the kidney without significant effect on glutamine synthesis (7).

VIII. Evidence that the γ-Glutamyl Cycle Functions in Vivo Utilizing Intracellular Glutathione and Extracellular Amino Acid

In considering the function of the γ-glutamyl cycle, it is notable that glutathione is found almost exclusively intracellularly, whereas γ-glutamyl transpeptidase is bound to the membranes of cells located at sites known to be extensively involved in transport. Since the transpeptidase is accessible to externally supplied substrates, it has been concluded that this enzyme is bound to the outer surface of the cell membrane. The evidence for this derives from several approaches. Thus, suspensions of lymphoid cells act on γ-glutamyl substrates that are impermeable to the cell membrane. Disruption of such cells by sonication does not lead to increased transpeptidase activity (32). In studies on a closed intact perfused rat kidney system, evidence was obtained, by analysis of the perfusate, that glutathione introduced into the system was rapidly broken down into its constituent amino acids (22,28). When glutathione is injected into animals it disappears rapidly; the kidney appears to be a major site of glutathione breakdown (15). γ-Glutamyl transpeptidase can be liberated from renal tubule cells by perfusion of the kidney with papain (21). Whereas these studies show that the transpeptidase is accessible to substrates outside the cell, they do not exclude the possibility that the enzyme is also accessible to the intracellular glutathione, and in fact the available data indicate that the transpeptidase acts significantly on intracellular glutathione. Since γ-glutamyl transpeptidase is the major, if not the only enzyme that degrades glutathione, it follows that intracellular gluthatione must be accessible to the transpeptidase in order to account for the significant turnover of intracellular glutathione. If the transpeptidase is located only on the outer surface of the membrane, then there must be a channel or carrier system of some sort to transport intracellular glutathione to the transpeptidase. Alternatively, the transpeptidase may also be located in the membrane at regions other than the external surface. In either case, the data strongly indicate that the transpeptidase can act on both intracellular and extracellular glutathione.

Consistent with the idea that intracellular glutathione is transported to the membrane-bound transpeptidase is the finding of substantial extracellular glutathione in a patient with γ-glutamyl transpeptidase deficiency. Schulman et al. (45) concluded that this patient's glutathionuria is secondary to his glutathionemia, and that glutathione is secreted by or leaks from renal cells into the glomerular filtrate. It is probable that the transport of intracellular glutathione to the plasma in this patient represents an aspect of a process that occurs normally to provide substrate to the membrane-bound enzyme. In the absence of the enzyme, glutathione leaks out of the cells into the plasma. Renal transpeptidase may also act on glutathione that enters the systemic circulation as a consequence of cell destruction. However, it seems unlikely that such glutathione could account for as much as 850 mg per day, the amount excreted in the urine by this patient. Glutathione is readily converted in blood plasma to glutathione disulfide (45), which is, compared to glutathione, a very poor substrate of the transpeptidase (52).

If membrane-bound γ-glutamyl transpeptidase interacts with intracellular glutathione and extracellular amino acid or peptide acceptors, it should be possible to demonstrate utilization of glutathione after administration of such acceptors. The finding that tissue levels of 5-oxoproline increase in the presence of L-2-imidazolidone-4-carboxylate and, that these levels increase more after administration of amino acids together with the inhibitor, is consistent with this hypothesis (64). Thus, administration of amino acids would be expected to increase the extent of transpeptidation, and the γ-glutamyl amino acids formed would become substrates of intracellular γ-glutamyl cyclotransferase. The finding that β-glutamyl-α-aminobutyrate, an inhibitor of γ-glutamyl cyclotransferase, inhibits the normal formation of 5-oxoproline and also its accumulation in the presence of L-2-imidazolidone-4-carboxylate strongly supports this interpretation.

When glycylglycine, an excellent acceptor substrate of transpeptidase, was administered to rats there was a marked decrease in tissue glutathione, an effect which was ascribed to the action of γ-glutamyl transpeptidase (38). Such a decrease in glutathione level was not seen after administration of equivalent doses of glycine or other amino acids. Presumably after giving small doses of amino acids the glutathione used for transpeptidation is rapidly regenerated by synthesis so that the glutathione level remains the same. In the experiments with glycylglycine, however, the very rapid transpeptidation reaction that occurs evidently utilizes glutathione at a faster rate than that of its resynthesis so that the glutathione level falls. It seemed possible that administration of a larger amount of amino acid might also increase the rate of transpeptidation to an extent which would deplete intracellular glutathione. To test this idea, mice were given relatively large doses of amino acids and the intracellular concentrations of glutathione and 5-oxoproline were determined. Dramatic decreases in the level of glutathione in the kidney were found after administration of L-methionine and certain other amino acids at a dose of 32 mmol per kg. Thus, after giving methionine, the level of renal glutathione decreased to about 50% of the control level in 30 min and to about 35% of the control in 1 h (8). In these experiments there was also a substantial increase (about 2-fold) in the 5-oxoproline level. The most probable explanation for these results is that the increased concentration of plasma amino acids leads to substantially increased transpeptidation, and that this reaction uses intracellular glutathione more rapidly than it can be synthesized. Such utilization of glutathione would be expected to lead to increased intracellular levels of γ-glutamyl amino acids; that this did indeed occur is reflected by the finding of increased levels of 5-oxoproline. It thus became of interest to determine the effects of inhibition of γ-glutamyl transpeptidase on these phenomena. One would expect that administration of a large dose of amino acid to an animal whose transpeptidase has been inhibited should not produce a marked decline in intracellular glutathione or an increase in 5-oxoproline formation. This expectation was fulfilled in recent studies (8). In mice in which γ-glutamyl transpeptidase was inhibited, administration of amino acids did not lead to a marked decrease of glutathione levels in the kidney nor to an increase in 5-oxoproline levels.

Injection into mice of prothionine sulfoximine (which inhibits glutathione synthesis) is also followed by very rapid depletion of glutathione; the rate of this decline seems to reflect the normal rate of glutathione utilization by γ-glutamyl transpeptidase (7). The rate at which the glutathione level falls after injection of prothionine sulfomixine was found to be decreased when an inhibitor of γ-glutamyl transpeptidase was also given. These findings indicate that the steady state levels of glutathione in the kidney result from a balance between the rate of glutathione synthesis and the rate of its utilization by transpeptidase. This balance may be markedly perturbed by inhibiting glutathione synthesis or by increasing its utilization by transpeptidation.

The observations summarized above are consistent with the conclusion that amino acids enter cells by the γ-glutamyl cycle pathway. However, more data are needed to show *directly* that γ-glutamyl amino acids are translocated. The findings thus far show a close relationship between increased extracellular amino acid concentration on the one hand, and decreased intracellular glutathione levels and increased 5-oxoproline levels, on the other. It is conceivable that the transpeptidase can act on intracellular amino acids as well as on intracellular glutathione. Thus, amino acids transported into the cell by a mechanism not involving the γ-glutamyl cycle might interact with transpeptidase intracellularly; this would also be expected to lead to decreased levels of glutathione and increased levels of 5-oxoproline. The physiological value or purpose of such an "intracellular γ-glutamyl cycle" is not immediately evident, although it is conceivable that such reactions could play a role in moving amino acids into (or out of) various intracellular compartments. Alternative possibilities include the idea that such a cycle is designed to produce cysteine, on stimulation by other

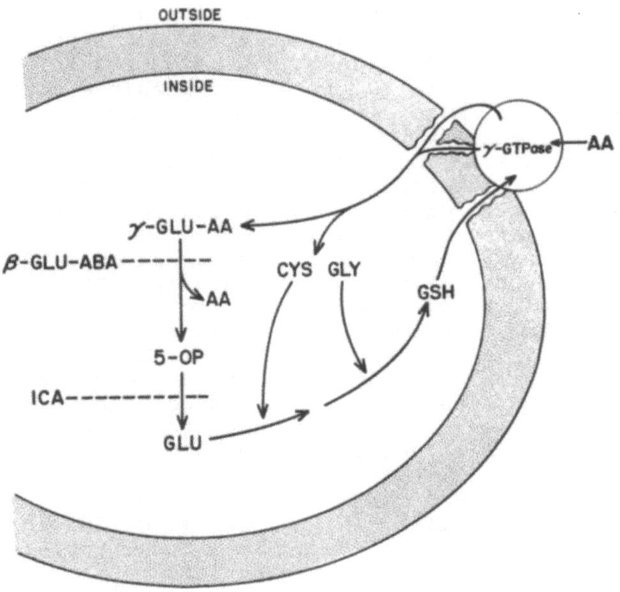

Fig. 3. Possible pathways of substrates to and from the membrane bound enzyme. γ-GTPase, γ-glutamyl transpeptidase; β-Glu-ABA, β-glutamyl-α-aminobutyric acid; ICA, 2-imidazolidone-4-carboxylic acid. Channels for transport of glutathione to the membrane-bound γ-glutamyl transpeptidase, and for transport of γ-glutamyl amino acid from outside the cell are consistent with the experimental data (see text). Glutathione and γ-glutamyl amino acids may be translocated by the same or a similar mechanism

amino acids, for protein synthesis, or that it is simply a futile cycle. Although the possibility of an intracellular γ-glutamyl cycle exists, its function would not be entirely clear; furthermore, the membranous localization of γ-glutamyl transpeptidase at various specific sites known to be involved in transport (nephron, intestinal villus, choroid plexus, etc.) suggests that the function of the γ-glutamyl cycle is related to a transport process.

The diagram given in Figure 3 reflects our current thinking about the γ-glutamyl cycle. We tentatively place the transpeptidase at the outer surface of the membrane (in agreement with current thought, but subject to the limitations discussed above). The enzyme is accessible to intracellular glutathione, which is in some manner transported to the enzyme; it is possible that a protein carrier may be involved. The transpeptidase interacts with amino acid and glutathione to form a γ-glutamyl amino acid which enters the cell. The orientation of the transpeptidase in the membrane may be such, as postulated previously (24), as to facilitate translocation, or alternatively there may be a separate pathway or channel for the translocation of the γ-glutamyl amino acid. That a translocation pathway for γ-glutamyl amino acids does exist is suggested by recent studies on the uptake of methionine sulfoximine by kidney (13). In these experiments, more methionine sulfoximine was found in the kidney after administration of γ-glutamyl methionine sulfoximine to mice than after giving methionine sulfoximine itself; in addition, γ-glutamyl methionine sulfoximine was found in the kidney. That γ-glutamyl methionine sulfoximine is well transported is consistent with the postulated function of the γ-glutamyl cycle.

The utilization of glutathione for mercapturic acid formation involves reactions of the γ-glutamyl cycle. Interaction of a foreign compound with glutathione, presumably an intracellular event catalyzed by the glutathione S-transferases (6), leads to formation of a S-substituted glutathione derivative, whose γ-glutamyl moiety is removed by γ-glutamyl transpeptidase. Presumably the S-substituted glutathione molecule is transported from within the cell to the membrane-bound enzyme to accomplish this enzymatic transformation. It seems significant that ligandins, cytoplasmic proteins that bind glutathione as well as various electrophilic compounds, also exhibit glutathione S-transferase activity. It is possible that ligandin not only serves as a catalyst for the formation of S-substituted glutathione, but that it also functions to transport these molecules to the transpeptidase. Ligandin may also be involved in the normal transport of glutathione to the transpeptidase. The glycine moiety of the S-substituted cysteinylglycine molecule is removed by the action of membrane-bound peptidase (18) or by intracellular peptidase activity. After removal of the γ-glutamyl and the C-terminal glycine moieties, the S-substituted cysteine molecule is acetylated by an intracellular system to form a mercapturic acid. The S-substituted glutathione molecule thus follows pathways (involving catalysis and translocation) which are closely parallel to those postulated for glutathione in the γ-glutamyl cycle. Further similarity is reflected by the observation that the activity of γ-glutamyl transpeptidase toward S-substituted glutathione is markedly increased by the presence of amino acids (52); in this reaction, the γ-glutamyl group is transferred to amino acid to form a γ-glutamyl amino acid. One may therefore expect that the occurrence of this process in vivo is followed by 5-oxoproline formation and utilization. It is interesting, in reviewing the parallel features of mercapturic acid formation and the γ-glutamyl cycle, to recall the comment of Boyland (2) that in the development of this pathway of detoxification, nature has adapted existing enzyme mechanisms for glutathione synthesis and utilization for purposes of detoxification.

Further work is required to establish the quantitative contribution of the γ-glutamyl cycle to the overall transport of amino acids, and to determine its relative activity with different amino acids. In relation to the latter, studies on the specificity of γ-glutamyl cyclotransferase, which indicate that the γ-glutamyl derivatives of glutamine, alanine, methionine, and a few others are the best substrates, might suggest that these free amino acids are the major cycle substrates; however, the occurrence of transpeptidation reactions between the γ-glutamyl derivatives of other amino acids and glutamine, and the formation of di-γ-glutamyl amino acids by transpeptidation offer possible pathways for other amino acids. The membrane-bound γ-glutamyl transpeptidase is a key enzyme in this series. We need to know more about its structure, mechanism of action, and orientation in the cell-membrane.

D. Conclusion

In summary, current studies provide substantial evidence that the γ-glutamyl cycle functions in vivo, and that it accounts for the very considerable turnover of intracellular glutathione. The steady-state levels of tissue glutathione reflect a balance between glutathione utilization and glutathione biosynthesis. When biosynthesis is stopped by administration of a specific inhibitor, renal glutathione levels decline rapidly, and the rate of this decline is slowed by simultaneous administration of an inhibitor of γ-glutamyl transpeptidase. Similarly, the glutathione level declined rapidly after administration of large doses of amino acids, and this decline is also slowed by inhibition of transpeptidase. Imidazolidone carboxylate, a competitive inhibitor of 5-oxoprolinase, produces 5-oxoproline accumulation in vivo. β-Glutamyl-α-aminobutyrate, which competitively inhibits γ-glutamyl cyclotransferase, decreases the physiological formation of 5-oxoproline and also its accumulation after administration of imidazolidone carboxylate. These effects reflect a major physiological connection between the function and metabolism of glutathione and the transport of amino acids. Although a direct demonstration of the translocation step is still required, the finding of markedly increased 5-oxoproline levels after amino acid administration (and prevention of this by inhibition of transpeptidase), and the observation that γ-glutamyl amino acids can be transported intact seem to offer strong support for the proposal that the cycle functions as *one* of the systems that mediate amino acid transport. None of these findings excludes the possibility that γ-glutamyl transpeptidase can also function in vivo to hydrolyze glutathione, S-substituted glutathione derivatives (as in mercapturic acid formation), or other γ-glutamyl compounds. The cycle also functions to store cysteine in the form of glutathione, and in the reversal of this process. Thus, the concept of the γ-glutamyl cycle continues to serve usefully in stimulating new experiments on amino acid translocation and glutathione metabolism.

References

1. Adamson, E.D., Szewczuk, A., Connell, G.E.: Purification and properties of γ-glutamyl cyclotransferase from pig liver. Canad. J. Biochem. 49, 218-226 (1971)
2. Boyland, E.: Mercapturic acid conjugation. Proc. 1st Int. Pharmacol. Meeting 1961, 16, 65-76 (1963)
3. Bridges, R.J., Griffith, O.W.: γ-glutamyl cyclotransferase-inhibition studies *in vitro* and *in vivo* using β-aminoglutaryl-L-α-aminobutyrate. Federation Proc. 37, 388 (1978)
4. Connell, G.E., Hanes, C.S.: Enzymic formation of pyrrolidone carboxylic acid from γ-glutamyl peptides. Nature (London) 177, 377-378 (1956)
5. Gass, J.D., Meister, A.: Computer analysis of the active site of glutamine synthetase. Biochemistry 9, 1380-1390 (1970)
6. Arias, I.M., Jakoby, W.B. (eds.): Glutathione; metabolism and function. Krok Foundation Series, Vol.6, New York: Raven Press 1976
7. Griffith, O.W., Anderson, M.E., Meister, A.: Inhibition of glutathione biosynthesis by prothionine sulfoximine (S-n-propyl-homocysteine sulfoximine), a selective inhibitor of γ-glutamylcysteine synthetase. J. Biol. Chem. 254 (1979)
8. Griffith, O.W., Bridges, R.J., Meister, A.: Evidence that the γ-glutamyl cycle functions *in vivo* using intracellular glutathione: Effects of amino acids and selective inhibition of enzymes. Proc. Natl. Acad. Sci. U.S., 75 (No.11) (1978)
9. Griffith, O.W., Meister, A.: Dependence of energy coupling on nucleotide base structure in the reaction catalyzed by 5-oxo-prolinase. Biochem. Biophys. Res. Commun. 70, 759-765 (1976)
10. Griffith, O.W., Meister, A.: Selective inhibition of γ-glutamyl-cycle enzyme by substrate analogs. Proc. Natl. Acad. Sci. U.S.A. 74, 3330-3334 (1977)
11. Griffith, O.W., Meister, A.: Differential inhibition of glutamine and γ-glutamylcysteine synthetases by α-alkyl analogs of methionine sulfoximine that induce convulsions. J. Biol. Chem. 253, 2333-2338 (1978)
12. Griffith, O.W., Meister, A.: unpublished
13. Griffith, O.W., Meister, A.: unpublished
14. Griffith, O.W., Van Der Werf, P., Meister, A.: Specificity and mechanism of action of 5-oxoprolinase, glutathione: metabolism and function. Arias, I.M., Jakoby, W.B. (eds.), Krok Foundation Symposium, Santa Barbara, California. June 1-3, 1975; pp.63-69. New York: Raven Press 1976

15. Hahn, R., Wendel, A., Flohé, L.: The fate of extracellular glutathione in the rat. Biochim. Biophys. Acta 539, 324-337 (1978)
16. Hanes, C.S., Hird, F.J.R., Isherwood, F.A.: Enzymatic transpeptidation reactions involving γ-glutamyl peptides and α-aminoacyl peptides. Biochem. J. 51, 25-35 (1952)
17. Heinz, E.: Transport of amino acids by animal cells. In: Metabolic pathways. Hokin, L.F. (ed.), 6, 455-501. New York: Academic Press 1972
18. Hughey, R.P., Rankin, B.A., Elce, J.S., Curthoys, N.P.: Specificity of a particulate rat renal peptidase and its localization along with other enzymes of mercapturic acid synthesis. Arch. Biochem. Biophys. 186, 211-217 (1978)
19. Kanazawa, A., Sano, I.: The distribution of γ-L-glutamyl-L-glutamine in mammalian tissues. J. Neurochem. 14, 596-598 (1967)
20. Konrad, P.N., Richards, F., Valentine, W.N., Paglia, D.E.: γ-glutamylcysteine synthetase deficiency. A cause of hereditary hemolytic anemia. N. Engl. J. Med. 286, 557-561 (1972)
21. Kuhlenschmidt, T., Curthoys, N.P.: Subcellular localization of rat kidney phosphate independent glutaminase. Arch. Biochem. Biophys. 167, 519-524 (1975)
22. Maack, T., Johnson, V., Tate, S.S., Meister, A.: Effects of amino acids on the function of the isolated perfused rat kidney. Federation Proc. 33, 305 (1974)
23. Majerus, P.W., Brauner, M.J., Smith, M.B., Minnick, V.J.: Glutathione synthesis in human erythrocytes. II. Purification and properties of the enzymes of glutathione biosynthesis. J. Clin. Invest. 50, 1637-1643 (1971)
24. Meister, A.: On the enzymology of amino acid transport. Science 180, 33-39 (1973)
25. Meister, A.: The γ-glutamyl cycle; diseases associated with specific enzymatic deficiencies. Ann. Int. Med. 81, 247-253 (1974)
26. Meister, A.: 5-oxoprolinuria (pyroglutamic aciduria) and other disorders of glutathione biosynthesis. In: The metabolic basis of inherited diseases, 4th ed. Stanbury, J.B., Wyngaarden, J.B., Frederickson, D.S. (eds.). Chap.16; pp. 328-336. New York: McGraw-Hill 1977
27. Meister, A., Bukenberger, M.W., Strassburger, M.: The optically specific enzymatic cyclization of D-glutamate. Biochem. Z. (Otto Warburg Festband) 338, 217-229 (1963)
28. Meister, A., Tate, S.S.: Glutathione and related γ-glutamyl compounds; biosynthesis and utilization. Ann. Rev. Biochem. 45, 559-604 (1976)
29. Meister, A., Tate, S.S., Ross, L.L.: Membrane-bound γ-glutamyl transpeptidase. In: The enzymes of biological membranes. Martinosi, A. (ed.). 3, 315-347. New York: Plenum 1976
30. Meister, A., Tate, S.S., Thompson, G.: On the function of the γ-glutamyl cycle in the transport of amino acids and peptides, pp. 123-143. CIBA Foundation Symposium on Peptide Transport and Hydrolysis, London; Nov. 1976 (1977)
31. Mooz, E. Dodd, Meister, A.: Tripeptide (glutathione) synthetase; purification properties, and mechanism of action. Biochemistry 6, 1722-1734 (1967)
32. Novogrodsky, A., Tate, S.S., Meister, A.: γ-glutamyl transpeptidase. A lymphoid cell-surface marker; relationship to blastogenesis, differentiation, and neoplasia. Proc. Natl. Acad. Sci. U.S.A. 73, 2414-2418 (1976)
33. Oppenheimer, L., Wellner, V.P., Chernov, N., Griffith, O.W.: Mapping of γ-glutamylcysteine binding site of glutathione synthetase; high specificity at the cysteine site relative to the glutamyl site. Federation Proc. 37, 848 (1978)
34. Orlowski, M., Meister, A.: The γ-glutamyl cycle; possible transport system for amino acids. Proc. Natl. Acad. Sci. U.S.A. 67, 1248-1255 (1970)
35. Orlowski, M., Meister, A.: Isolation of highly-purified γ-glutamylcysteine synthetase from rat kidney. Biochemistry 10, 372-380 (1971)
36. Orlowski, M., Meister, A.: γ-glutamyl cyclotransferase; distribution, isozymic forms, and specificity. J. Biol. Chem. 248, 2836-2844 (1973)
37. Orlowski, M., Richman, P., Meister, A.: Isolation and properties of γ-L-glutamylcyclotransferase from human brain. Biochemistry 8, 1048-1055 (1969)
38. Palekar, A.G., Tate, S.S., Meister, A.: Decrease in glutathione levels of kidney and liver after injection of methionine sulfoximine into rats. Biochem. Biophys. Res. Commun. 62, 651-657 (1975)
39. Pardee, A.B.: Membrane transport proteins. Science 162, 632-637 (1968)

40. Revel, J.P., Ball, E.G.: The reaction of glutathione with amino acids and related compounds as catalyzed by γ-glutamyl transpeptidase. J. Biol. Chem. 234, 577-582 (1959)
41. Richards, F., Cooper, M.R., Pearce, L.A., Cowan, R.J., Spurr, C.L.: Familial spinocerebellar degeneration, hemolytic anemia, and glutathione deficiency. Arch. Int. Med. 134, 534-537 (1974)
42. Richman, P., Meister, A.: Regulation of γ-glutamyl-cysteine synthetase by non-allosteric feedback inhibition by glutathione. J. Biol. Chem. 250, 1422-1426 (1975)
43. Richman, P.G., Orlowski, M., Meister, A.: Inhibition of γ-glutamylcysteine synthetase by L-methionine-S-sulfoximine. J. Biol. Chem. 248, 6684-6690 (1973)
44. Ronzio, R.A., Meister, A.: Phosphorylation of methionine sulfoximine by glutamine synthetase. Proc. Natl. Acad. Sci. U.S.A. 59, 164-170 (1968)
45. Schulman, J.D., Goodman, S.I., Mace, J.W., Patrick, A.D., Tietze, F., Butler, E.J.: Glutathionuria: inborn error of metabolism due to tissue deficiency of γ-glutamyl transpeptidase. Biochem. Biophys. Res. Commun. 65, 68-74 (1975)
46. Sekura, R., Meister, A.: Glutathione turnover in the kidney; considerations relating to the γ-glutamyl cycle and the transport of amino acids. Proc. Natl. Acad. Sci. U.S.A. 71, 2404-2409 (1974)
47. Sekura, R., Meister, A.: γ-glutamylcysteine synthetase: further purification, "half of the sites" reactivity, subunits, and specificity. J. Biol. Chem. 252, 2599-2605 (1977)
48. Sekura, R., Van Der Werf, P., Meister, A.: Mechanism and significance of the mammalian pathway for elimination of D-glutamate; inhibition of glutathione synthesis by D-glutamate. Biochem. Biophys. Res. Commun. 71, 11-18 (1976)
49. Taniguchi, N., Meister, A.: γ-glutamyl cyclotransferase from rat kidney; sulfhydryl groups and isolation of a stable form of the enzyme. J. Biol. Chem. 253, 1799-1806 (1976)
50. Tate, S.S.: Interaction of γ-glutamyl transpeptidase with S-acyl derivatives of glutathione. FEBS Lett. 54, 319-322 (1975)
51. Tate, S.S., Meister, A.: Stimulation of the hydrolytic activity and decrease of the transpeptidase activity of γ-glutamyl transpeptidase by maleate; identity of a rat kidney maleate-stimulated glutaminase and γ-glutamyl transpeptidase. Proc. Natl. Acad. Sci. U.S.A. 71, 3329-3333 (1974)
52. Tate, S.S., Meister, A.: Interaction of γ-glutamyl transpeptidase with amino acids, dipeptides, and derivatives and analogs of glutathione. J. Biol. Chem. 249, 7593-7602 (1974)
53. Tate, S.S., Meister, A.: Identity of maleate-stimulated "glutaminase" with γ-glutamyl transpeptidase in rat kidney. J. Biol. Chem. 250, 4619-4624 (1975)
54. Tate, S.S., Meister, A.: Subunit structure and isozymic forms of γ-glutamyl transpeptidase. Proc. Natl. Acad. Sci. U.S.A. 73, 2599-2603 (1976)
55. Tate, S.S., Meister, A.: Affinity labeling of γ-glutamyl transpeptidase and location of the γ-glutamyl binding site on the light subunit. Proc. Natl. Acad. Sci. U.S.A. 74, 931-935 (1977)
56. Tate, S.S., Meister, A.: Serine-borate complex as a transition-state inhibitor of γ-glutamyl transpeptidase. Proc. Natl. Acad. Sci. U.S.A. 75, 4806-4809 (1978)
57. Tate, S.S., Ross, M.E.: Human kidney γ-glutamyl transpeptidase. J. Biol. Chem. 252, 6042-6045 (1977)
58. Thompson, G.A., Meister, A.: Utilization of L-cystine by the γ-glutamyl transpeptidase-γ-glutamyl cyclotransferase pathway. Proc. Natl. Acad. Sci. U.S.A. 72, 1985-1988 (1975)
59. Thompson, G.A., Meister, A.: Hydrolysis and transfer reactions catalyzed by γ-glutamyl transpeptidase; evidence for separate substrate sites and for high affinity of L-cystine. Biochem. Biophys. Res. Commun. 71, 32-36 (1976)
60. Thompson, G.A., Meister, A.: Interrelationships between the binding sites for amino acids, dipeptides, and γ-glutamyl donors in γ-glutamyl transpeptidase. J. Biol. Chem. 252, 6792-6797 (1977)
61. Thompson, G.A., Meister, A.: Hippurate, a possible regulator of γ-glutamyl transpeptidase. Proc. Natl. Acad. Sci. U.S.A. 76 (1979)
62. Van Der Werf, P., Griffith, O., Meister, A.: 5-oxo-prolinase (L-pyroglutamate hydrolase); purification and catalytic properties. J. Biol. Chem. 250, 6686-6692 (1975)

63. Van Der Werf, P., Orlowski, M., Meister, A.: Enzymatic conversion of 5-oxo-L-proline (L-pyrrolidone carboxylate) to L-glutamate coupled with ATP cleavage to ADP: a reaction in the γ-glutamyl cycle. Proc. Natl. Acad. Sci. U.S.A. 68, 2982-2985 (1971)

64. Van Der Werf, P., Stephani, R.A., Meister, A.: Accumulation of 5-oxoproline in mouse tissues after inhibition of 5-oxoprolinase and administration of amino acids; evidence for function of the γ-glutamyl cycle. Proc. Natl. Acad. Sci. U.S.A. 71, 1026-1029 (1974)

65. Van Der Werf, P., Meister, A.: The metabolic formation and utilization of 5-oxo-L-proline (L-pyroglutamate, L-pyrrolidone carboxylate). Adv. Enzymol. 43, 519-556 (1975)

66. Wellner, V.P., Sekura, R., Meister, A., Larsson, A.: Glutathione synthetase deficiency, an inborn error of metabolism involving the γ-glutamyl cycle in patients with 5-oxoprolinuria (pyroglutamic aciduria). Proc. Natl. Acad. Sci. U.S.A. 71, 2505-2509 (1974)

67. Wendel, A., Schaich, E., Weber, U., Flohé, L.: Isolierung und Molekulargewichtsbestimmung der γ-L-Glutamyl-L-Cystein: Glycin-Ligase. Hoppe-Seyler's Z. Physiol. Chem. 343, 514-522 (1972)

68. Wilson, H., Cannon, R.K.: The glutamic acid-pyrrolidonecarboxylic acid system. J. Biol. Chem. 119, 309-331 (1937)

69. Woodward, G.E., Reinhart, F.E.: The effect of pH on the formation of pyrrolidone carboxylic acid and glutamic acid during enzymatic hydrolysis of glutathione by rat kidney extract. J. Biol. Chem. 145, 471-480 (1942)

Topology and Function of Renal γ-Glutamyl Transpeptidase

S. SILBERNAGL, W. PFALLER, H. HEINLE, and A. WENDEL

Summary

The fate of glutathione and some of its analogs and derivatives was studied by micro-perfusion of single rat nephrons in vivo et in situ. The accessibility and localization of the enzyme in the brush border membrane of the proximal convolution was investigated by means of electron microscopy following immunochemical labeling of the enzyme in vivo.

^{14}C-glycine-labeled glutathione and ophthalmic acid were degraded in the proximal convolution with half-lives of 1.5 s and 0.8 s, respectively. S-methyl-GSH and S-sulfobromophthalein-GSH as representatives of a simple and a bulky mercapturic acid precursor either labeled in glutamate or in the sulfur substituent were degraded with half-lives of 3.5 s to S-methyl-cysteine and S-sulfobromophthaleinecysteine. In the case of the recognized impermeable substituent sulfobromophthaleine, no radioactivity was found in the perfusate when glutamate was originally labeled, whereas with labeled sulfobromophthaleine a total recovery of radioactivity was observed.

Monospecific antibodies were raised in the rabbit against rat kidney γ-GT. The γ-globulin fraction was microperfused through the rat nephron in vivo. After washing, anti-rabbit immunoglobulin from the goat coupled with ferritin was perfused.

Electron micrographs showed ferritin granules all along the brush border membrane in a distance where the glycocalix ends. Both the apical and basal part of the microvilli were stained.

This study demonstrates that renal γ-GT is an extrinsic brush border membrane protein involved in the extracellular, intraluminal hydrolysis of glutathione and a final step of mercapturic acid synthesis.

A. Introduction

γ-Glutamyl transferase (E.C.2.3.2.2) is abundant in kidney and many other organs (22). Most of the renal enzyme is localized in the brush border fraction (14,15). Histochemical studies with artificial chromogenic substrates showed that the enzyme activity is intimately associated with the brush border membrane (2,24). Experiments using kidney cones (9) and single microdissected parts of the rat nephron (17) demonstrated that the enzyme exhibits maximal activity in proximal straight part of the tubule.

In vitro the enzyme catalyzes at least four reactions:

1. Hydrolysis of GSH and S-substituted derivatives.
2. Transpeptidation of the glutamyl moiety of GSH to various acceptors, among which are certain amino acids.

3. Autotranspeptidation between two GSH molecules forming cysteinyl-glycine+γ-glu-GSH.
4. Hydrolysis of glutamine to glutamate + NH_3.

Recently, the enzyme has been shown to be identical with the phosphate-independent glutaminase (8,30).

Despite detailed knowledge on the molecular (4,18,21,31,32,37) and kinetic (11,12,20) properties, the physiological functions of renal γ-GT are still open to discussion. Several functions have been suggested: hydrolysis of glutathione (3), mediation of renal amino acid transport (6,24), renal uptake or secretion of γ-glutamyl compounds (13,25), oligopeptide group translocation (26), participation in secretory processes (5), splitting of extracellular GSH and GSSG (33), involvement in mercapturic acid synthesis (7,12,19,32) and, finally, homoiostasis of the renal pH by the enzyme's glutaminase activity sensitive to metabolic alkalosis or acidosis (8).

The aim of the present study was to confirm our suggestion (33) that glutathione is split extracellularly in the tubular lumen, to extend this claim on the degradation of mercapturic acid precursors, and to show that the enzyme is appropriately localized in the brush-border membrane to achieve this function. Therefore, glutathione as well as some of its analogs and derivatives were microperfused through single rat nephrons in vivo et in situ. Their rates of degradation and the products were determined. Furthermore, by the same technique the enzyme was labeled immunochemically within the tubule and visualized by means of electron microscopy.

B. Materials and Methods

Male Wistar rats were kept without food but with free access to water for 15 h prior to the experiments. After anesthesia with 120 mg/kg Inactin (Promonta, Hamburg), the left kidney was exposed through a subcostal incision, immobilized in a lucite cap and covered with mineral oil maintained at 37.5°C. Sections of single proximal tubules or the straight tubule plus Henle's loop were continuously perfused with "equilibrium solutions" (29) containing 115 mEq/l Na^+, 5 mEq/l K^+, and 4 mEq/l Ca^{2+}, buffered by Tris-(hydroxy-methyl)-methyl-2-aminoethane-sulfonic acid (TES) pH = 7.4. The osmolarity of these solutions was adjusted to 290-300 mOsmol/l by addition of mannitol. Further additions are given in the legends. The perfusion rate was 20 nl/min. Tritium-labeled inulin, perfused simultaneously, served as internal standard. At the end of each experiment the perfused segment was filled with latex. The kidney was removed and partly macerated with 6 N HCl. The tubular casts were cut out and their length was measured under the microscope. The radioactivity of the perfusate was determined by liquid scintillation counting corrected for the inulin value and compared with the initial activity. In addition, the perfusate was analyzed for peptides and amino acids using an ultra-sensitive automated amino acid analysis system with fluorescence detection after reaction with o-phthaldialdehyde (29). For further details of the technique see (10).

(^{14}-C)-glutamate-labeled GSH was synthesized by means of a continuous enzyme-reactor: γ-glutamylcysteine-synthetase (specific activity 0.25 U/mg) and glutathione synthetase (0.8 U/mg) were co-immobilized on glass-fiber.

To a 10 x 100 mm column, a total volume of 10 ml was applied at 37°C, which contained 0.3 mM ^{14}C-glutamate (10 mCi/mmol), 0.6 mM cysteine, 5 mM dithiothreitol, 5 mM glycine, 5 mM ATP, 10 mM Mg SO_4, 50 mM KCl and 100 mM triethanolamine-HCl buffer pH = 7.4. After 7 h of recycling 80% GSH had been synthesized. The products were purified on a 1 x 100 cm Biogel P 2 column (Biorad, München) in water.

Labeled S-methyl-glutathione was synthesized as follows:

1. 1.25 µmol glutamate-labeled GSH reacted with 30 µmol methyliodide (Fluka) at pH = 9.0 for 2 h at 25°C in the dark. Purification was performed on a 1 x 100 mm Biogel P 2 column in 50 mM formic acid/ pyridine pH = 6.0.
2. 4.6 µmol (^{14}C)-methyliodide dissolved in methanol (5 µCi/µmol) reacted with 1 µmol GSH (Waldhof-Pharma) under the same conditions. HJ and unreacted methyliodide + solvent were removed under vacuum. Both compounds were lyophilized.

The adduct of GSH with bromosulfaleine (BSP-GSH) was synthesized according to the following procedure:

3. 1.15 µmol ^{35}S-bromosulfaleine (10 mCi/mmol) reacted with 1.15 µmol GSH for 2 h at 37°C. The conjugate was precipitated with acetone and dried (35).
4. 1.25 µmol (^{14}C-glutamate)-GSH reacted with 10 µmol bromosulfalein (Fluka) for 2 h. The product was analogously isolated.

The purity of the compounds (1) through (4) was controlled using descending thin layer chromatography on cellulose in butanol: acetic acid: water = 4 : 1 : 1. All compounds as localized by ninhydrin staining and radioscanning of the chromatogram were radiochemically pure.

All other labeled compounds were obtained from Amersham-Buchler. γ-glutamyl transpeptidase was purified to apparent homogeneity from rat kidney by a new procedure (34) using hollow fiber dialysis and concentration, hydrophobic chromatography on phenyl sepharose (Pharmacia), and gel filtration on sephacryl S 200 (Pharmacia).

The purification was 735-fold with a yield of 18% at a final specific activity of 840 U/mg. Rabbits were immunized against γ-GT with the pure enzyme protein by monthly injections of 0.2 and 0.4 mg of protein in Freund's adjuvans. Bovine serum albumin (5 x crystallized, Serva) was injected as control. The globulin fraction of the rabbit serum was precipitated by $(NH_4)_2SO_4$, dissolved in H_2O and dialyzed against 50 mM potassium phosphate/150 mM NaCl pH = 7.2. With this fraction a single precipitation line was observed against pure γ-GT in the immuno-diffusion test according to Ouchterlony.

5.5 mg/ml anti-γ-GT-globulin from immunized or BSA-immunized rabbits were microperfused in equilibrium solution for 5 min through single rat proximal convolutions. Subsequently, the tubule was washed by a 1-min perfusion with equilibrium solution. Then, a 0.1 mg/ml equilibrium solution of anti-rabbit-immunoglobulin raised from the goat, coupled with ferritin (Miles-Yeda), was perfused for 4 min, again washed for 1 min, and finally 1% glutardialdehyde buffered with Tyrode solution (21) was perfused for 5 min.

The kidney was removed, the area of the perfused tubule was cut out and fixed for 1 h in 1% glutardialdehyde. The further treatment was: 30 min washing in Tyrode buffer according to (21), 1 h in Tyrode buffer without glucose containing 1% OsO_4, 30 min washing in Tyrode; dehydration by increasing acetone concentrations; embedding in Durcupan-ACM

Fluka). Thin sections of 600 Å thickness were cut on a Reichert OmU 3 ultra-microtome and contrasted by $Bi(NO_3)_3$ (1). Electron micrographs were taken using a Jeol 100 C electron microscope at 100 kV.

C. Results

Figure 1 shows that GSH is rapidly degraded in the proximal tubule and removed from the perfusate. From these data and the perfusate flow a half-life of 1.5 s is calculated for GSH. The fact that cysteinylglycine and glycine are found as labeled split products indicates that it is the tripeptide's γ-glutamyl bond which is cleaved off first.

Thus, on a few millimeters within the tubule, the GSH molecule is effectively degraded. This is even more pronounced after passage of Henle's loop, where no intact tripeptide but little cysteinylglycine plus glycine leave the early distal tubule (Fig. 1 right hand side). With ophthalmic acid, an even faster degradation is observed (Fig. 2, solid circles) which corresponds to an estimated half-life of 0.8 s. In this case, the perfusate was analyzed for the concentration of (cold) glutamate. It turned out that glutamate transiently accumulates and is then reabsorbed (Fig.2, solid line designated as control). This, however, is not the case if glutamate reabsorption is competitively inhibited by aspartate (36): then the amount of free glutamate determined in the perfusate exactly corresponds to the amount of ophthalmic acid which has been degraded. Evidently, the degrading activity proceeds independently of transport of glutamate. After perfusion of the loop, ophthalmic acid was completely split, but the major part of free glutamate was still found in the perfusate (Fig. 2, right hand side).

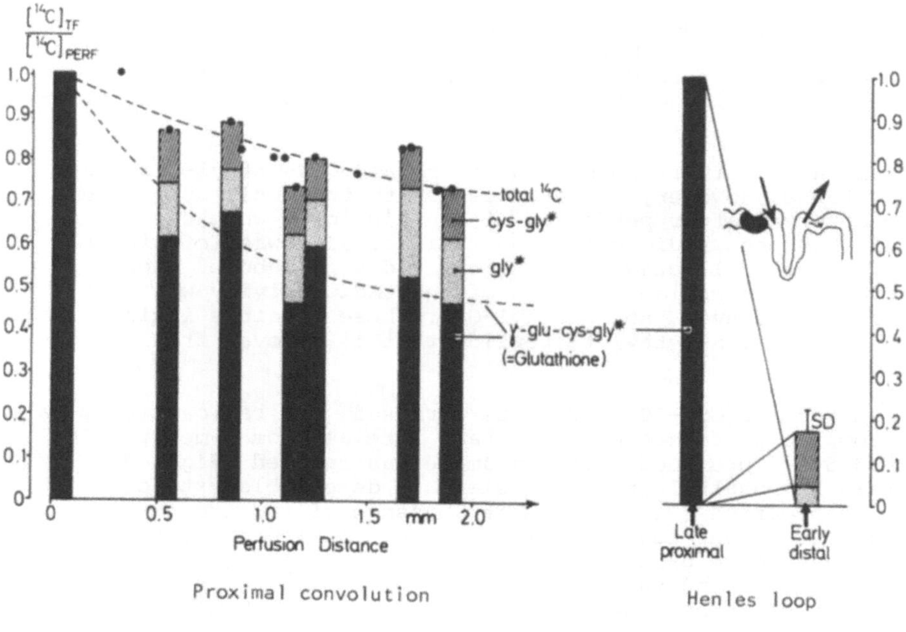

Fig. 1. Microperfusion of 1 mM [^{14}C-glycine]-GSH through single proximal convolutions and Henle loops of the rat kidney in vivo et in situ. Fractional recovery of the labeled compounds is plotted against perfusion distance in mm. The bars represent single experiments with the same kidney. •, total radioactivity

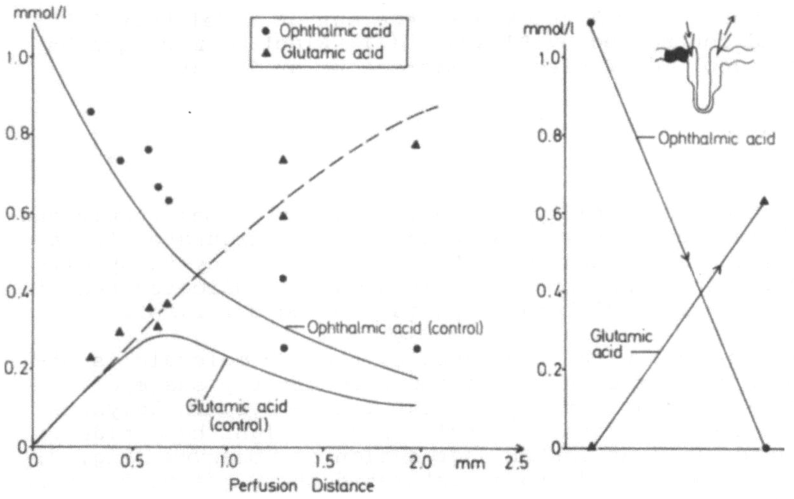

Fig. 2. Microperfusion of 1.1 mM unlabeled ophthalmic acid through proximal convolutions or Henle loops including pars recta in the presence of 20 mM aspartate. ●, concentration of ophthalmic acid. Intratubular concentration of glutamate in the presence of aspartate (▲) and in the absence of aspartate (control, solid line, n=9) during perfusion with ophthalmic acid

Two further experiments were performed in order to establish that also mercapturic acid precursors are degraded, and to show that this takes place in the tubular lumen before the compounds enter the tubular wall cell. Two S-substituted derivatives were microperfused through the rat nephron. Radioactive S-methyl-GSH, either labeled in the methyl group (Fig. 3A) or in the γ-glutamyl moiety (Fig. 3B) disappeared from the tubular lumen with half-lives of 3.5 s. The position of the label did not significantly influence the rate of absorption of the radioactivity.

The residual radioactivity consisted of unchanged S-methyl-GSH plus S-methyl-cysteine. However, differences due to the position of the label were observed after perfusion of Henle's loop: when the γ-glutamyl moiety had been radioactive, no radioactivity was left in the perfusate that was collected from the early distal tubule. When the S-methyl group had been labeled, 37% of the radioactivity was left, which contained, however, no S-methyl-GSH. These findings indicate that it is not the intact S-methyl-tripeptide that is removed from the tubular fluid.

When the radioactive GSH-BSP adduct is perfused, the radioactivity is absorbed from the perfusate with the same rate as from S-methyl-GSH (half-life 3.5 s), provided that glutamate was labeled (Fig. 4B). If the sulfur in bromosulfalein was labeled, no detectable extraction of radioactivity occurred (Fig. 4A). After passage of the loop, this different behaviour of one and the same compound labeled at different positions is even more pronounced. This means that the glutamyl moiety is first cleaved off and then reabsorbed. The remaining cysteinyl derivative cannot be transported since it bears the impermeable ligand BSP.

There seems to be no doubt that the only enzyme capable of cleaving the γ-glutamyl peptide bond of GSH, ophthalmic acid and the two chem-

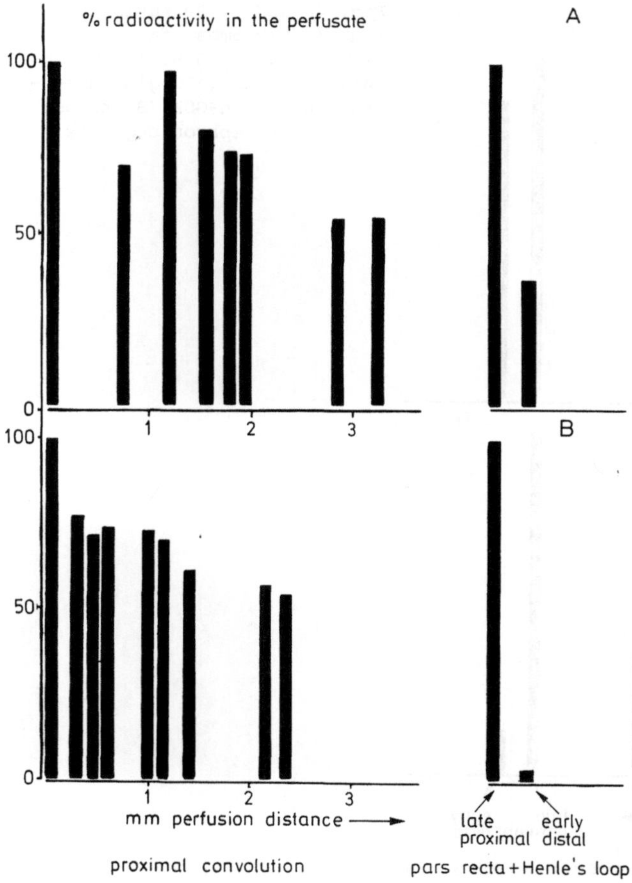

Fig. 3A and B. Perfusion of 2 mM (^{14}C-methyl) S-methyl GSH, 5 mCi/mmol (A) and 1 mM (^{14}C-glutamate)-GSH, 10 mCi/mmol (B) through single rat nephron segments

ically very different S-substituted derivatives of GSH is γ-glutamyltranspeptidase. If the view is correct that this enzyme is able to split these compounds so rapidly when they pass by in the tubular fluid, then it must be localized as far as possible on the outside of the brush border membrane in order to bind and hydrolyze its substrates. Figure 5 shows as a result of an indirect immunochemical labeling that this is indeed the case. The ferritin granules indicate the presence of a γ-GT-anti-γ GT complex which bound the ferritin-coupled anti-rabbit-immuno-globulin. It can be seen that these granules are localized all along the brush border membrane on the apical as well as on the basal parts of the microvilli and sometimes within phagocytic vacuoles. The distance to the middle of the membranes ranges from 150-200 Å, i.e., it amounts to the thickness of the glycocalix. When the same experiment was performed with perfusion of the γ-globulin fraction of rabbit immunized with bovine serum albumin instead of rat γ-GT, very few granules with an absolutely random distribution were observed. Thus γ-GT is protruding out of the brush border membrane and is even accessible for a large antibody molecule.

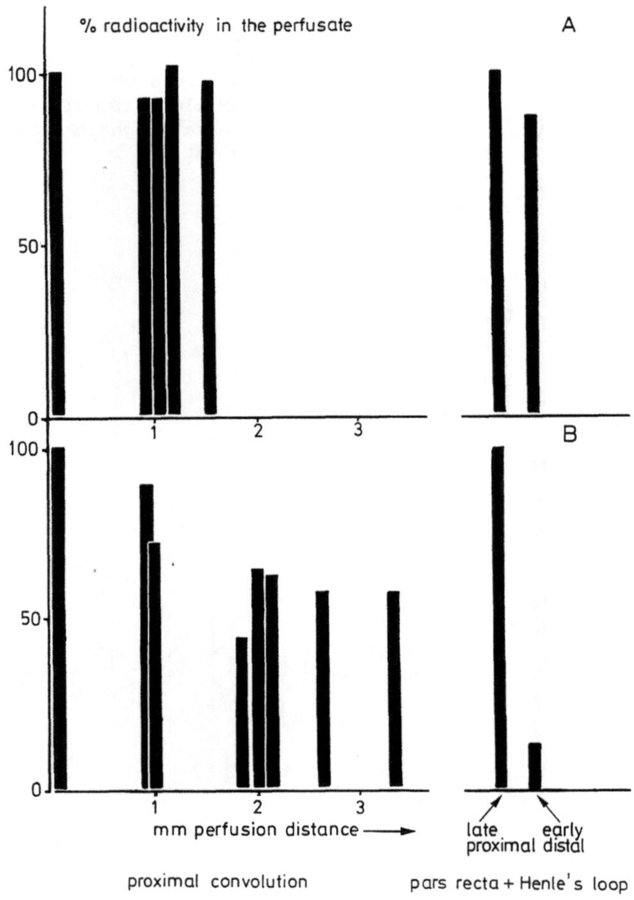

Fig. 4A and B. Perfusion of 1 mM (^{35}S-bromosulfalein)-bromosulfalein-GSH, 10 mCi/mmol (A), and 1 mM (^{14}C-glutamate)-bromosulfalein-GSH (B) through single rat nephron segments

D. Discussion

The present study demonstrates that under in vivo conditions renal γ-glutamyl transpeptidase catalyzes the degradation of extracellular glutathione which reaches the tubule following glomerular filtration. With a GSH release rate of 12 nmol/min per g rat liver (cf. article of Bartoli, Häberle and Sies, this volume) this function of γ-GT gains quantitative significance. This seems to apply also to humans: a γ-GT-deficient patient exhibited glutathionuria and excreted 850 mg glutathione per day in the urine (28). In addition, our investigation shows that γ-GT catalyzes in vivo a final step in mercapturic acid synthesis by extracellular splitting of the γ-glutamyl moiety of S-substituted GSH derivatives. These functions may be quantitatively even more significant under emergency situations when the liver excretes biotransformation products of xenobiotica. The ultrastructural evidence shows that γ-GT is an extrinsic brush border membrane protein protruding out of the glycoprotein layer of the brush border membrane at least to such an extent that it is recognized by an antibody molecule.

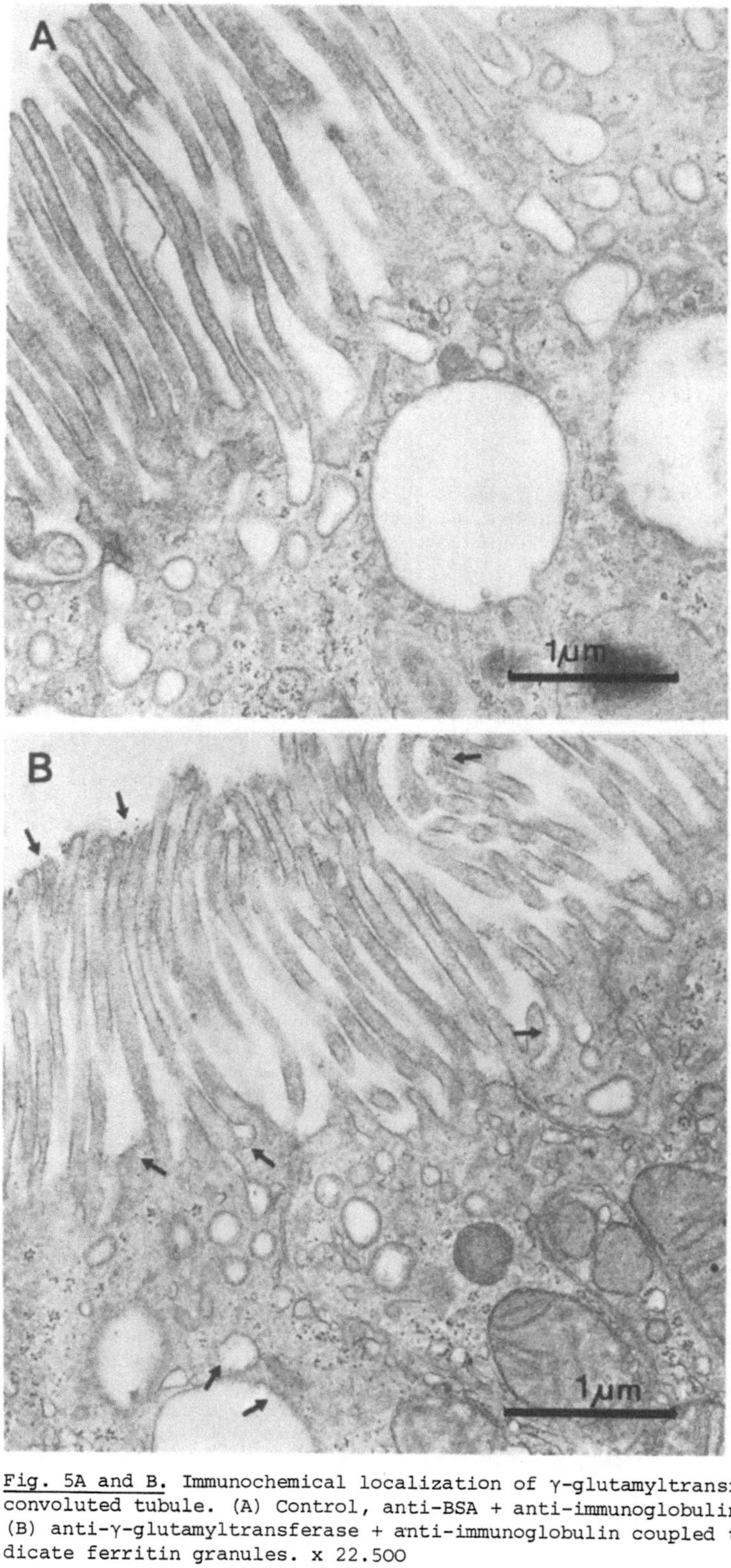

Fig. 5A and B. Immunochemical localization of γ-glutamyltransferase in a single rat convoluted tubule. (A) Control, anti-BSA + anti-immunoglobulin coupled to ferritin; (B) anti-γ-glutamyltransferase + anti-immunoglobulin coupled to ferritin. Arrows indicate ferritin granules. x 22.500

The distant localization at the glycocalix border suggests that γ-GT is not a transmembrane protein with a second binding site for intracellular GSH. The finding that transport of glutamate can be blocked while hydrolysis of the γ-glutamyl residue of ophthalmic acid is unaffected disagrees with a participation of the enzyme in γ-glutamyl peptide or amino acid transport (24). This statement is only valid for the investigated structures of the kidney, i.e., proximal convoluted or straight tubule and Henle's loop, and transport or accessibility in the direction glomerular filtrate → tubular cell. The possibility is open that the inside of the brush border or the two sides of the basal-lateral membranes may also contain low but significant amounts of transpeptidase. It is conceivable that the basal side of the tubule cell or other parts of the kidney account for the observed accumulation of γ-glutamylpeptides within the whole organ after i.v. administration of such compounds (22,25).

Acknowledgments. This work was financially supported by the Deutsche Forschungsgemeinschaft (Grant We 686/4). Thanks go to Mrs. G. Vetter, Innsbruck and to Dipl.-Biochem. W. Rollinger, Tübingen for helpful collaboration.

References

1. Ainsworth, S.K., Karnovsky, M.J.: An ultrastructural staining method for enhancing the size and electron opacity of ferritin in thin sections. J. Histochem. Cytochem. 20, 225-229 (1972)
2. Albert, Z., Orlowski, M., Szewczuk, A.: Histochemical demonstration of γ-glutamyl transpeptidase. Nature (London) 191, 767-768 (1961)
3. Binkley, F.: Metabolism of glutathione. Nature (London) 167, 888-889 (1951)
4. Binkley, F.: Purification and properties of renal glutathionase. J. Biol. Chem. 236, 1075-1082 (1961)
5. Binkley, F., Wiesemann, M.L.: Glutathione and γ-glutamyl transferase in secretory processes. Life Sci. 17, 1359-1362 (1976)
6. Bodnaryk, R.: Membrane-bound γ-glutamyl transpeptidase. Evidence that it is a component of the amino acid site of certain neutral amino acid transport systems. Can. J. Biochem. 50, 524-528 (1972)
7. Boyland, E., Chasseaud, L.F.: The role of glutathione and glutathione-S-transferases in mercapturic acid biosynthesis. Adv. Enzymol. 32, 173-291 (1969)
8. Curthoys, N.P., Kuhlenschmidt, T.: Phosphate-independent glutaminase from rat kidney. Partial purification and identity with glutamyl transpeptidase. J. Biol. Chem. 250, 2099-2105 (1975)
9. Curthoys, N.P., Lowry, O.H.: The distribution of glutaminase isoenzymes in the various structures of the nephron in normal, acidotic, and alkalotic rat kidney. J. Biol. Chem. 248, 162-168 (1973)
10. Deetjen, P., Silbernagl, S.: Some new developments in continuous microperfusion technique. Yale J. Biol. Med. 45, 301-306 (1972)
12. Elce, J.S., Bryson, J., Mc Girr, L.G.: γ-Glutamyl transpeptidase of rat kidney. Some properties and kinetic constants. Can. J. Biochem. 52, 33-41 (1973)
13. Fonteles, M.C., Pillion, D.J., Jeske, A.H., Leibach, F.H.: Extraction of glutathione by isolated perfused rabbit kidney. J. Surg. Res. 21, 169-174 (1976)
14. George, S.G., Kenny, A.J.: Studies on the enzymology of purified preparations of brush border from rabbit kidney. Biochem. J. 134, 43-57 (1973)
15. Glossmann, H., Neville, D.M.: γ-Glutamyl transferase in kidney brush border membranes. FEBS Lett. 19, 340-344 (1972)
16. Heinle, H., Sawatzki, G., Wendel, A.: Glutathionbiosynthese VII: Die Biosynthese von Glutathion in Humanerythrocyten. Hoppe-Seyl. Z. Physiol. Chem. 357, 1451-1458 (1976)

17. Heinle, H., Wendel, A., Schmidt, U.: The activities of the key enzymes of the γ-glutamyl cycle in microdissected segments of the rat nephron. FEBS Lett. 73, 220-224 (1977)
18. Hughey, R.P., Curthoys, N.P.: Comparison of the size and physical properties of γ-glutamyl transpeptidase purified from rat kidney following solubilization with papain or with triton X-100. J. Biol. Chem. 251, 7863-7870 (1976)
19. Hughey, R.P., Rankin, B., Elce, J.S., Curthoys, N.P.: Mercapturic acid synthesis I. Regional and subcellular localization of enzyme activities within rat kidney. Arch. Biochem. Biophys. 186, 211-217 (1978)
20. London, J.W., Shaw, L.M., Fetterolf, D., Garfinkel, D.: Determination of the mechanism and kinetic constants for hog kidney γ-glutamyltransferase. Biochem. J. 157, 609-617 (1976)
21. Maunsbach, A.B.: The influence of different fixation and fixation methods on the ultrastructure of rat kidney proximal tubule cells II: Effect of varying osmolarity, ionic strength, buffer system and fixative concentration of glutaraldehyde solutions. J. Ultrastruct. Res. 15, 283-289 (1966)
22. Meister, A., Tate, S.S., Ross, L.L.: Membrane-bound γ-glutamyl transpeptidase. In: The enzymes of biological membranes, Vol.3, 315-347 (1976)
23. Miller, S.P., Awasti, Y.C., Srivastava, S.K.: Studies of human kidney γ-glutamyl transpeptidase. J. Biol. Chem. 251, 2271-2278 (1976)
24. Orlowski, M., Meister, A.: The γ-glutamyl cycle: a possible transport system for amino acids. Proc. Natl. Acad. Sci. U.S.A. 67, 1248-1255 (1970)
25. Orlowski, M., Wilk, S.: Synthesis of ophthalmic acid in liver and kidney in vivo. Eur. J. Biochem. 170, 415-421 (1978)
26. Prusiner, S., Doak, C.W., Kirk, G.: A novel mechanism for group translocation: substrate-product utilization by γ-glutamyl transpeptidase in peptide and amino acid transport. J. Cell. Physiol. 89, 853-864 (1976)
27. Rutenberg, A.M., Kim, H., Fischbein, J.W., Hanker, J.S., Wasserkrug, H.L., Seligman, A.M.: Histochemical and ultrastructural demonstration of γ-glutamyl transpeptidase activity. J. Histochem. Cytochem. 17, 517-526 (1969)
28. Schulman, J.D., Goodman, S.I., Mace, J.W., Patrick, A.D., Tietze, F., Butler, E.J.: Glutathionuria: inborn error of metabolism due to tissue deficiency of gamma-glutamyl transpeptidase. Biochem. Biophys. Res. Commun. 65, 68-74 (1975)
29. Silbernagl, S., Bayliss, E.: unpublished (1977)
30. Tate, S.S., Meister, A.: Identity of maleate-stimulated glutaminase with γ-glutamyl transpeptidase in rat kidney. J. Biol. Chem. 250, 4619-4624 (1975)
31. Tate, S.S., Meister, A.: Subunit structure and isozymic forms of γ-glutamyltranspeptidase. Proc. Natl. Acad. Sci. U.S.A. 73, 2599-2603 (1976)
32. Thompson, G.A., Meister, A.: Interrelationships between the binding sites for amino acids, dipeptides, and γ-glutamyl donors in γ-glutamyl transpeptidase. J. Biol. Chem. 252, 6792-6798 (1977)
33. Wendel, A., Hahn, R., Guder, W.G.: On the role of γ-glutamyl transferase in renal tubular amino acid reabsorption. Curr. Probl. Clin. Biochem. 6, 426-436 (1976)
34. Wendel, A., Rollinger, W.: unpublished (1978)
35. Whelan, G., Hoch, J., Combes, B.: A direct assessment of the importance of conjugation for biliary transport of sulfobromophthalein sodium. J. Lab. Clin. Med. 75, 542-557 (1970)
36. Young, J.A., Freedman, B.S.: Renal tubular transport of amino acids. Clin. Chem. 17, 245-266 (1971)
37. Zelazo, W., Orlowski, M.: γ-Glutamyl transpeptidase of sheep-kidney cortex. Isolation, catalytic properties and dissociation into two polypeptide chains. J. Biochem. 61, 147-155 (1976)

The Membrane Association and Physiological Function of Rat Renal γ-Glutamyltranspeptidase

N. P. CURTHOYS, R. P. HUGHEY, and P. J. COYLE

Summary

The interaction of γ-glutamyltranspeptidase with membranes was studied by characterizing the association and orientation of this enzyme within lecithin vesicles. It was found that only γ-glutamyltranspeptidase purified following solubilization with Triton X-100 can associate with single-layered [^{14}C] lecithin vesicles. This was established by comparing enzyme activity and radioactivity profiles during Sepharose 4B chromatography and isopycnic sucrose gradient centrifugation. The enzyme-vesicle complex exhibits a density corresponding to that of a single enzyme molecule bound to a single vesicle. γ-Glutamyltranspeptidase purified following solubilization with papain does not bind to vesicles. In addition, papain treatment of vesicles containing the Triton-purified transpeptidase results in the release of 95% of the transpeptidase activity without release of internally trapped [^{3}H] sucrose. The released transpeptidase is chromatographically identical to the papain-purified transpeptidase. γ-Glutamyltranspeptidase activity associated with both native membranes and with lecithin vesicles exhibits a temperature-induced transition in its energy of activation. In contrast, the proteolytic and detergent solubilized forms of the enzyme exhibit a single energy of activation over the entire temperature range. These results suggest that γ-glutamyltranspeptidase binding to vesicles is due to a papain-sensitive sequence of amino acids and that the enzyme-vesicle complex closely approximates the interaction and orientation of γ-glutamyltranspeptidase with brush border membranes. These results are inconsistent with the hypothesis that this enzyme participates in amino acid transport via the γ-glutamyl cycle, but instead supports the proposal that the principal function of γ-glutamyl-transpeptidase is the extracellular degradation of glutathione.

A. Introduction

γ-Glutamyltranspeptidase is a broad specificity transferase which catalyzes the transfer of γ-glutamyl groups from a variety of γ-glutamyl peptides, including glutathione, to a wide range of amino acid and peptide acceptors (Tate and Meister, 1974). This lack of specificity and its localization in membranes principally involved in absorption or secretion have led to the proposal that this enzyme participates in amino acid transport via the γ-glutamyl cycle (Orlowski and Meister, 1970; Meister, 1973). But the ability of water to act as an acceptor suggests that γ-glutamyltranspeptidase participates in the degradation of extracellular glutathione (Hahn et al., 1978; Elce and Broxmeyer, 1976) and the biosynthesis of mercapturic acids (Hughey et al., 1978).

We report here the characterization of the interaction between the papain- and Triton-purified rat renal γ-glutamyltranspeptidase and an artificial membrane system of single-layered lecithin vesicles. Additional evidence for the membrane attachment of transpeptidase by a peptide anchor is produced by the finding that only the Triton-purified transpeptidase can associate with lecithin vesicles and that this association mimics the orientation and interaction exhibited by γ-glutamyltranspeptidase in native membrane.

B. Methods

Rat renal microsomes (Curthoys and Weiss, 1974) and Triton- and papain-purified forms of γ-glutamyltranspeptidase (Hughey and Curthoys, 1976) were prepared as described previously. Transpeptidase activity was determined using γ-glutamyl-p-nitroanilide as the substrate (Tate and Meister, 1974). Arrhenius plots were prepared by determining γ-glutamyltranspeptidase activity of various forms of the enzyme at temperature varying from 5°-40°C.

Lecithin vesicles were prepared by a slight modification of the procedure of Brunner et al. (1976). Egg lecithin (10 mg) in hexane and dipalmitoyl [^{14}C] lecithin (15 µg) in ethanol:toluene (1:1) were dried under a stream of nitrogen. Samples of [^{3}H] sucrose or γ-glutamyltranspeptidase (100 µg) were dried by lyophilization or acetone precipitation (Hughey and Curthoys, 1976), respectively. The dried materials were resuspended in 1 ml of vesicle buffer (0.1 M NaCl, 0.1 mM EDTA, 0.02% NaN_3 and 10 mM Tris-Cl, pH 7.2) containing 2% sodium cholate. After an overnight incubation at room temperature, the solution was applied to a Sephadex G-50 column (23 x 1.5 cm) in order to remove the sodium cholate. The void volume fractions were pooled and used for further analysis.

C. Results

The papain- and Triton-purified forms of γ-glutamyltranspeptidase were denatured and dissociated into individual subunits by treatment with 6 M urea (Tate and Meister, 1976). The individual subunits were isolated by gel filtration and analyzed by polyacrylamide gel electrophoresis in the presence of sodium lauryl sulfate (Laemmli, 1970). As shown in Figure 1, only the smaller subunits of the two forms of transpeptidase appear to be identical. The larger subunit of the Triton-purified transpeptidase migrates slower and is therefore larger than the corresponding subunit from the papain-purified transpeptidase.

Vesicles prepared from egg- and dipalmitoyl [^{14}C] lecithin produced the expected elution profile upon Sepharose 4B chromatography (Fig. 2A). Multi-layered vesicles elute in the void volume (fraction 19) and single-layered vesicles (MW ~ 4.6 x 10^6) are included in the column and elute as a peak near fraction 30. The Triton-purified transpeptidase aggregates in the absence of detergents or vesicles (MW ~ 1 x 10^6) and elutes from the Sepharose 4B column as a sharp peak at fraction 34. In contrast, the papain-purified enzyme (MW < 10^6) elutes as a sharp peak late in the column profile and is well separated from the vesicles. Therefore, the individual elution profiles are sufficiently different for Sepharose 4B chromatography to be a sensitive indicator of binding of transpeptidase to lecithin vesicles.

Chromatography of [^{14}C] lecithin vesicles prepared in the presence of the papain-purified transpeptidase produces a radioactivity and enzyme activity profile (Fig. 2B) which is nearly identical to that produced

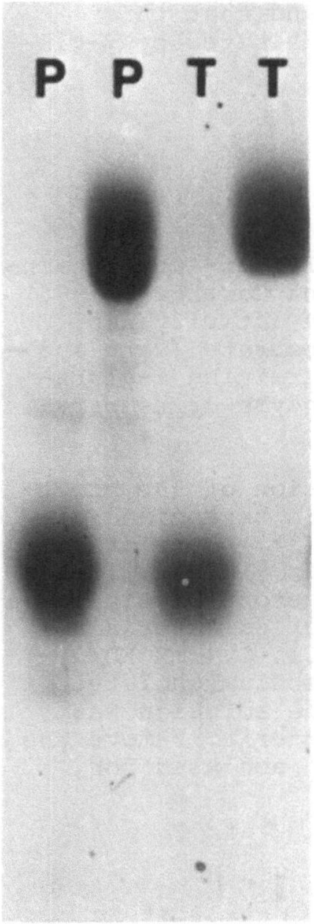

Fig. 1. Sodium lauryl sulfate polyacrylamide gel electrophoresis of subunits isolated from papain-purified (P) and Triton-purified (T) γ-glutamyltranspeptidase

by chromatography of either one alone. In contrast, chromatography of a sample of [^{14}C] lecithin vesicles prepared in the presence of the Triton-purified transpeptidase produces coincident peaks of enzyme activity and radioactivity (Fig. 2C). Following treatment of an identical sample with papain, 95% of the transpeptidase activity is released from the vesicles and elutes from the Sepharose 4B column (Fig. 2D) in the fractions characteristic of the papain-purified transpeptidase.

Proof of transpeptidase binding to vesicles requires determination of the density of the lecithin-protein complex. During isopycnic centrifugation all of the transpeptidase activity floats into a gradient and bands as a symmetrical peak at 6.5% sucrose (Fig. 3). This corresponds to the density calculated for a single molecule of transpeptidase (MW 87,000) bound to a single-layered lecithin vesicle (MW 4.6×10^6). From the proportion of [^{14}C] lecithin associated with the transpeptidase activity, one can estimate that 10%-20% of the vesicles bind γ-glutamyltranspeptidase. Centrifugation of an identical sample pretreated with papain yields a profile which is very similar to that obtained with [^{14}C] lecithin vesicles prepared in the presence of papain-purified transpeptidase. In either case, only 5% of the transpeptidase activity bands at 6.5% sucrose, and the rest of the activity remains at the bottom of the gradient.

As shown in Figure 4, isopycnic centrifugation of vesicles formed in the presence of [^{3}H] sucrose produces profiles similar to those obtained with [^{14}C] lecithin vesicles. Treatment of an identical sample with papain releases 95% of the enzyme activity without releasing the

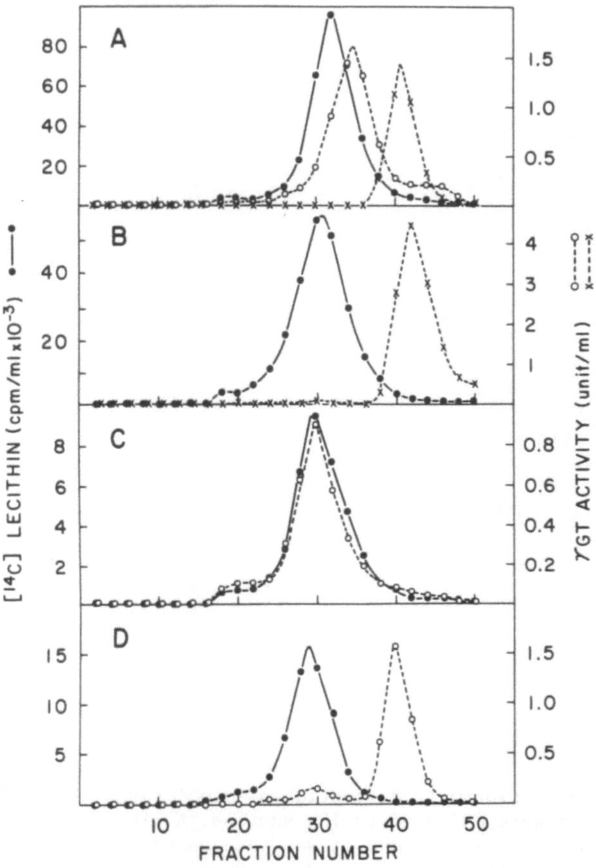

Fig. 2A-D. Association of γ-glutamyltanspeptidase with [^{14}C] lecithin vesicles. Aliquots of either vesicles, or transpeptidase were applied to a Sepharose 4B column. (A) Vesicles prepared from lecithin in the absence of protein (closed circles), Triton-purified (open circles) or papain-purified (crosses) transpeptidase chromatographed separately. (B) Vesicles prepared from lecithin in the presence of papain-purified transpeptidase. (C) Vesicles prepared from lecithin and Triton-purified transpeptidase. (D) Vesicles, prepared as in (C), and preincubated 1 h at 37°C with papain

[^{3}H] sucrose. Sepharose 4B chromatography of these [^{3}H] sucrose-loaded vesicles produces profiles similar to those shown in Figure 2C and D (data not shown).

As a further characterization of the similarity of transpeptidase binding to lecithin vesicles and to native membranes, the effect of temperature on enzymatic activity was investigated (Fig. 5). Only enzyme activity associated with either native or artificial membranes exhibits a temperature-induced transition in the energy of activation. γ-Glutamyltranspeptidase in native membranes exhibits a transition near 28°C, while enzyme in egg-lecithin vesicles exhibits a transition near 18°C. Triton-purified transpeptidase in Triton X-100 and papain-purified transpeptidase exhibit the same energy of activation over the entire temperature range.

D. Discussion

In the experiments reported here, the interaction of γ-glutamyltranspeptidase with membranes was characterized by determining the association and orientation of transpeptidase within lecithin vesicles. The observed coincidence of elution of Triton-purified transpeptidase ac-

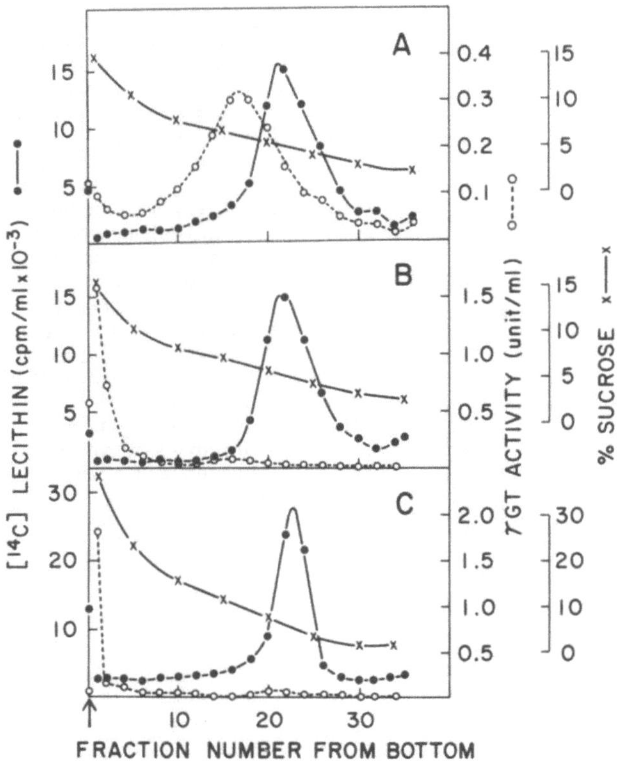

Fig. 3A-C. Isopycnic centrifugation of γ-glutamyltranspeptidase associated with [^{14}C] lecithin vesicles. Samples of vesicles were adjusted to 15% sucrose. A linear sucrose gradient (10%-2%) was layered on top of the sample and it was then centrifuged for 16 h at 40,000 rpm in an SW 41 rotor. Gradients were pumped off from the bottom and fractions were assayed for radioactivity (closed circles), enzyme activity (open circles) or refractive index (crosses). The empty tubes were washed with one fraction volume of vesicle buffer containing 2% cholate and the wash (arrow at origin) was assayed as described above. (A) Vesicles were prepared from lecithin and Triton-purified transpeptidase. (B) Sample prepared as in (A) and preincubated for 1 h at 37° with papain. (C) Vesicles prepared from lecithin and papain-purified transpeptidase

tivity and [^{14}C] lecithin during Sepharose 4B chromatography is consistent with the binding of this form of enzyme to vesicles. The other possible interpretation of these results is that aggregates of Triton-purified transpeptidase bind lecithin, and thus shift the enzyme activity towards the void volume. The isopycnic sucrose gradient analysis clearly distinguishes between these two possibilities. The lecithin-protein complex is shown to have a density corresponding to that expected for a single molecule of transpeptidase bound to a single-layered vesicle. Aggregates of enzyme which bind lecithin would be expected to exhibit a much greater density.

The ability of only the Triton-purified γ-glutamyltranspeptidase to bind to lecithin vesicles and the ability of papain to destroy this association while producing a smaller unaggregated form of the enzyme is consistent with the idea that a papain-sensitive peptide is responsible for binding of the transpeptidase to membranes. The experiments

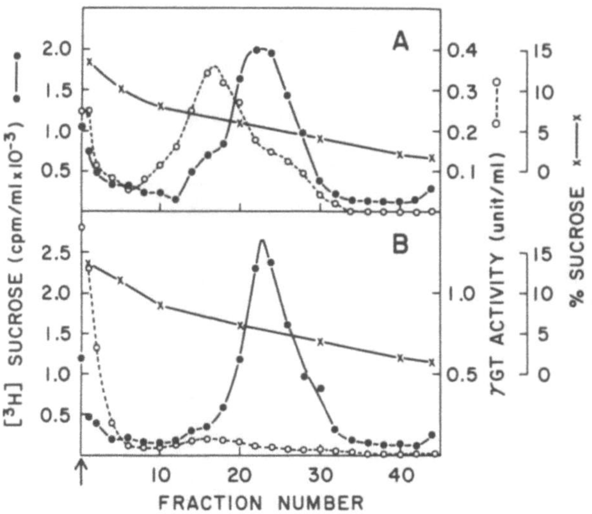

Fig. 4A and B. Retention of [^{3}H] sucrose in lecithin vesicles following papain solubilization of Triton-purified γ-glutamyltranspeptidase. Vesicles were prepared using lecithin, Triton-purified transpeptidase and 50 μCi [^{3}H] sucrose. Samples were analyzed by isopycnic sucrose gradient centrifugation, as described in the legend to Figure 3, following incubation for 1 h at 37°C either without additions (A) or with papain (B)

with [^{3}H] sucrose loaded vesicles indicates that papain solubilization of enzyme occurs without disrupting the bilayer or internal volume of the vesicle. Therefore, the papain-sensitive peptide orients γ-glutamyltranspeptidase primarily on the outside surface of the vesicles.

γ-Glutamyltranspeptidase associated with brush border membranes is completely resistant to inactivation by papain. The Triton-purified transpeptidase associated with lecithin vesicles exhibits a similar resistance to papain. However, in both cases papain treatment results in release of greater than 95% of the transpeptidase activity. This similar susceptibility to limited proteolysis is a good indication of retention of native structure in a solubilized or reconstituted membrane protein. Further evidence that the Triton-purified transpeptidase-vesicle complex approximates the in vivo membrane association of this enzyme comes from the observed membrane-dependent temperature-induced transition in its energy of activation. Therefore, the papain-sensitive peptide is probably also responsible for the association and orientation of γ-glutamyltranspeptidase within the rat renal brush border membrane.

γ-Glutamyltranspeptidase is composed of two distinct subunits, both of which are glycoproteins. Sodium lauryl sulfate polyacrylamide gel electrophoresis of the isolated subunits suggests that the papain-sensitive peptide is propably associated with the larger subunit. Studies with a radioactive affinity label indicate that the γ-glutamyl binding site is associated with the smaller subunit (Inŏue et al., 1977a; Tate and Meister, 1977). Due to its high content of carbohydrate, its extent of solubilization with papain, and its solubility in aqueous solutions, the active site is probably restricted to one side of the cell and never crosses the membrane.

Recent experiments by Kuhlenschmidt and Curthoys (1975) have shown that papain treatment of intact kidney cells results in solubilization of a large proportion of the γ-glutamyltranspeptidase activity without causing release of intracellular enzymes. In addition, Hahn et al. (1978) have shown that isolated kidney tubules degrade extracellular glutathione at rates several orders of magnitude greater than the rate of renal turnover of intracellular glutathione. These observations suggest

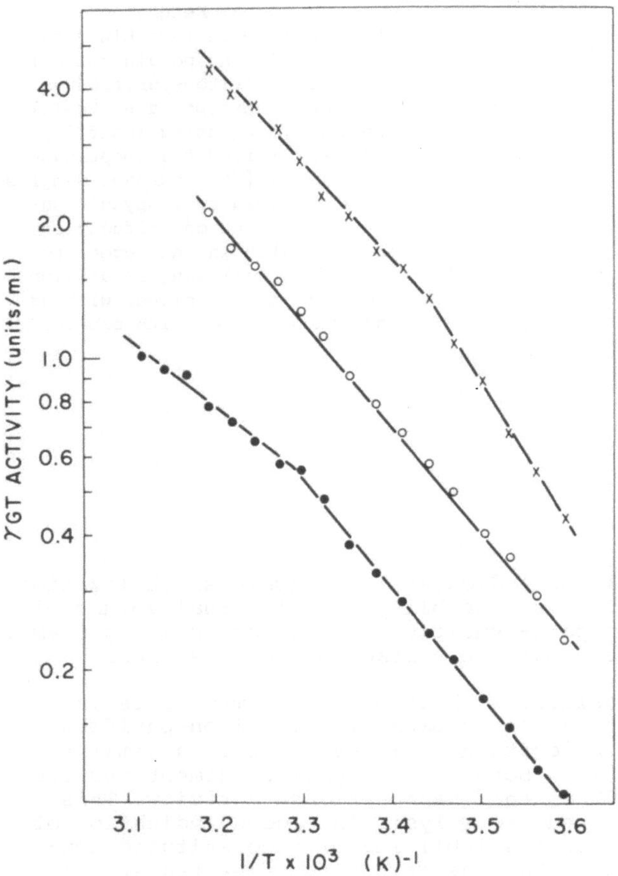

Fig. 5. Arrhenius plots of γ-glutamyltranspeptidase activity. Enzyme activities of isolated rat renal microsomes (closed circles) and of Triton-purified transpeptidase in 1% Triton X-100 (open circles) or in association with lecithin vesicles (crosses) were determined at temperature from 5°C to 40°C

that γ-glutamyltranspeptidase activity is primarily extracellular. Elce and Broxmeyer (1976) developed a highly sensitive assay for γ-glutamyltranspeptidase using [^{14}C]-glutamate labeled glutathione and reported that at physiological concentrations of methionine, the principal reaction catalyzed was hydrolysis and not transfer. In addition, a patient who lacks γ-glutamyltranspeptidase activity, has been reported to exhibit glutathionemia and glutathionuria, but not aminoaciduria (Schulman et al., 1975).

The cumulative evidence suggests that extracellular glutathione is the principal substrate for γ-glutamyltranspeptidase. Both reduced and oxidized glutathione are effectively hydrolyzed by γ-glutamyltranspeptidase (T. McIntyre and N.P. Curthoys, unpubl.). The concentration of glutathione in serum is normally very low (Tietze, 1969), but it has been shown that erythrocytes (Srivastava and Beutler, 1969a,b), lens (Srivastava and Beutler, 1969c) and liver (Sies et al., 1972; Sies and Summer, 1975) release oxidized glutathione at an increased rate in response to oxidant stress. The low serum levels of glutathione may be due to its efficient degradation by γ-glutamyltranspeptidase. Therefore, glutathione turnover may represent a process of intracellular synthesis, secretion and extracellular degradation.

Acknowledgments. This research was supported in part by research grants AM 16651 from the National Institute of Arthritis, Metabolism and Digestive Diseases and PCM 7681534 from the National Science Foundation.

References

Brunner, J., Skrabel, P., Hauser, H.: Biochim. Biophys. Acta 455, 322-331 (1976)
Curthoys, N.P., Weiss, R.F.: J. Biol. Chem. 249, 3261-3266 (1974)
Elce, J.S., Broxmeyer, B.: Biochem. J. 153, 223-232 (1976)
Hahn, R., Wendel, A., Flohé, L.: Biochim. Biophys. Acta 539, 324-337 (1978)
Hughey, R.P., Curthoys, N.P.: J. Biol. Chem. 251, 7863-7870 (1976)
Hughey, R.P., Rankin, B.B., Elce, J.S., Curthoys, N.P.: Arch. Biochem. Biophys. 186, 211-217 (1978)
Inoue, M., Horiuchi, S., Morino, Y.: Eur. J. Biochem. 73, 335-342 (1977)
Kuhlenschmidt, T., Curthoys, N.P.: Arch. Biochem. Biophys. 167, 519-524 (1975)
Laemmli, U.K.: Nature (London) 227, 680-685 (1970)
Meister, A.: Science 180, 33-39 (1973)
Orlowski, M., Meister, A.: Proc. Natl. Acad. Sci. U.S.A. 67, 1248-1255 (1970)
Schulman, J.D., Goodman, S.I., Mace, J.W., Patrick, A.D., Tietze, F., Butler, E.J.: Biochem. Biophys. Res. Commun. 65, 68-74 (1975)
Sies, H., Gerstenecker, C., Menzel, H., Flohé, L.: FEBS Lett. 27, 171-175 (1972)
Sies, H., Summer, K.: Eur. J. Biochem. 57, 503-512 (1975)
Srivastava, S.K., Beutler, E.: Biochem. J. 114, 833-837 (1969a)
Srivastava, S.K., Beutler, E.: J. Biol. Chem. 244, 9-16 (1969b)
Srivastava, S.K., Beutler, E.: Biochem. J. 112, 421-425 (1969c)
Tate, S.S., Meister, A.: J. Biol. Chem. 249, 7593-7602 (1974)
Tate, S.S., Meister, A.: Proc. Natl. Acad. Sci. U.S.A. 73, 2599-2603 (1976)
Tate, S.S., Meister, A.: Proc. Natl. Acad. Sci. U.S.A. 74, 931-935 (1977)
Tietze, F.: Anal. Biochem. 27, 502-510 (1969)

Studies on the Active Site of γ-Glutamyltransferase

J.S. ELCE

Summary

γ-Glutamyltransferase (E.C.2.3.2.2) from rat kidney catalyzes the hydrolysis of glutathione via a covalent γ-glutamyl-enzyme intermediate. The identity of some active site amino acid residues and of the residue which is covalently attached to the γ-glutamyl group was investigated by means of chemical modification. The enzyme was not inhibited by p-hydroxymercuribenzoate or by phenylmethylsulfonyl fluoride, implying the absence of active site cysteine or serine residues, but a cysteine residue was nonetheless later shown to be close to the active site. Inhibition by carbodiimide suggested the presence of an active site COOH group, and by trinitrobenzene sulfonic acid suggested that a lysine residue was important. Reaction of the active site with [^{14}C] iodoacetamide, in conjunction with N-acetylimidazole treatment, led to labeling both of a COOH group and of a cysteine residue. Treatment of the enzyme with N-acetylimidazole modified a lysine residue, and inhibited the enzyme but did not block formation of the γ-glutamyl-enzyme intermediate. This lysine residue when not acetylated appears to protect the cysteine residue from reaction with iodoacetamide. The γ-glutamyl group in the covalent enzyme intermediate is stable to performid acid, and to 1 M hydroxylamine in 6 M urea at pH 9, but is removed slowly at pH 12, suggesting that the γ-glutamyl residue must be attached to a second lysine residue by an amide linkage. This isopeptide linkage is known from studies on fibrin cross-linking, and attempts are under way to isolate γ-[^{14}C] glutamyl-ε-lysine from the γ-glutamyl-enzyme intermediate.

A. Introduction

Our previous kinetic studies have shown that the enzyme γ-glutamyltransferase (γGT) (E.C.2.3.2.2) has a branched non-sequential catalytic mechanism, and that a γ-[^{14}C] glutamyl-enzyme intermediate can be isolated (9). We wish to report here the results of further work aimed at characterizing the chemical linkage of the γ-glutamyl group to the enzyme; in addition, several other amino acid residues at or close to the active site have been identified.

Earlier efforts to identify active site residues were not conclusive (10,17,29), mainly because less pure samples of the enzyme were then available. The work was also made easier for us by recent work on chemical modification of proteins (20) and by the availability of [^{14}C] Glu-GSH (9).

B. Methods

The enzyme was obtained by papain treatment of rat kidney microsomes followed by chromatography on Sephadex G-200 (8) and then on Concanavalin-A Sepharose-4B (31). The final product was about 90% pure, judged from its specific activity and γ-glutamyl binding capacity. The activity was assayed routinely with γ-glutamyl-p-nitroanilide (γ-Glu-pNA) (21,28), and one unit of enzyme activity is defined as that amount of enzyme which catalyzes the formation of 1 μmol of p-nitroaniline per min. From the report of Tate and Ross (33) it may be calculated that 1 nanomol of γGT possesses about 70 units of activity. The activity of γGT was also assayed by measurement of the hydrolysis of [^{14}C] Glu-GSH (9). At 125 μM GSH and pH 8.0, 1 unit of γGT hydrolyses about 65 nanomol of GSH per min.

High voltage electrophoresis was carried out at 60 volts/cm for 0.75 h on Whatman 3 MM paper in acetic acid/formic acid/water (4:1:45, by vol), pH 2.0. Descending paper chromatography was performed in n-butanol/acetic acid/water (60:15:25, by vol) on Whatman No. 1 paper for 24 h.

C. Results and Discussion

The combination of serine and borate is known to inhibit γGT competitively (24,29) and was used in the present work to protect the active site of the enzyme from modification.

The time and concentration dependence of γGT inhibition were studied for a number of reagents, and the results are summarized in Table 1.

Table 1. Inhibition of γ-glutamyltransferase

Reagent[a]	Time (min)	Inhibition[b] (%)	Comments
1 mM p-hydroxymercuri-benzoate		None	
0.1 mM diisopropyl phospho-fluoridate		None	
1 mM phenylmethylsulfonyl-fluoride		None	
6.7 mM pyridoxal P. ± $NaBH_4$		None	
0.5 M H_2O_2	30	50	Not prevented by
	120	50	serine-borate.
0.2 M carbodiimide	15	50	Not reversed by
	100	80	1 M NH_2OH.
1 mM trinitrobenzene sulfonic	40	50	
acid	200	84	
N-acetylimidazole 12 mM	40	58	Not prevented by
31 mM	40	79	serine-borate or
56 mM	40	91	by substrates.
Iodoacetamide 2.5 mM	40	35	Prevented by
7.5 mM	40	68	serine-borate.
25 mM	40	95	

[a]Reactions were performed with 0.5 μM γGT in 10 mM Tris-HCl or phosphate buffers, pH 6.5-8.5.

[b]The enzyme was assayed with γ-Glu-p-nitroanilide (28).

The failure of mercuric compounds to inhibit γGT was well known (29, 34) and implies the absence of a cysteine from the active site, but in the experiments with iodoacetamide described below, a cysteine residue was nonetheless found to be carboxymethylated and close to the active site. The inhibition observed with H_2O_2 (Table 1) was probably a result of general oxidation of the enzyme, rather than suggesting a specific role of methionine (27). No other evidence for a methionine residue at the active site was obtained, although it has recently been reported in abstract (32) that a methionine is present at the active site of fibroblast γGT.

The inhibition observed with 1-ethyl-3(3-dimethylaminopropyl) carbodiimide (Table 1) implied that a COOH group was present at the active site, and this suggestion was confirmed by iodoacetamide treatment. Reaction of a tyrosine residue with the carbodiimide was not significant in the inhibition, as shown by treatment of inhibited γGT with NH_2OH (4,5,14).

Whereas pyridoxal phosphate (3,18) failed to inhibit γGT, it was found that trinitrobenzenesulfonic acid did cause an inhibition (Table 1), suggesting the importance of a lysine residue.

D. N-acetylimidazole

With 0.1 M N-acetylimidazole at pH 7.0, about 95% inhibition of γGT was achieved in 2 h. (Katunuma et al. (17) described the effect of N-acetylimidazole on rat kidney phosphate-independent glutaminase, which is identical to γGT (8).) No protection against N-acetylimidazole was afforded by 20 mM serine-borate, by 1 mM GSH or by 20 mM glycylglycine. Although the reagent can acetylate histidine, tyrosine or lysine side chains (25), in the present case the activity of the inhibited enzyme was not increased by incubation at pH 4.7, which argues against histidine acetylation (2), by incubation at pH 8.8, and by incubation with 0.9 M NH_2OH at pH 7.5, conditions which cause loss of tyrosine-O-acetates (2). Reaction with N-acetylimidazole therefore appears to cause acetylation of a lysine residue, or residues.

E. Iodoacetamide Labeling

The most useful results were obtained with iodoacetamide and N-acetylimidazole treatments of the enzyme, combined in various sequences. Szewczuk and Connell (29) were unable to identify the residues modified by iodoacetamide, but Connell and Adamson (6) mentioned the possibility that a COOH group on the enzyme was involved.

An active site-labeling of the enzyme with [^{14}C] iodoacetamide, taking advantage of the effects of N-acetylimidazole treatment, was achieved as follows. γGT (610 units, about 9-10 nanomol) was incubated at 20°C for 3 h with 25 mM serine-borate and 20 mM non-radioactive iodoacetamide, in order to modify any susceptible non-active site residues. Excess β-mercaptoethanol was added and the enzyme was dialyzed at 4°C for 24 h against 1 mM Tris-HCl, pH 8.0. The dialyzed enzyme, which retained its full activity, was divided into two equal portions. Portion I was untreated, and portion II was exposed to 0.1 M N-acetylimidazole at 4°C overnight, and was then found to be 95% inhibited. Both portions were precipitated with acetone and the precipitates were treated with 10 mM [^{14}C] iodoacetamide (5 µCi/ mol) in the dark at 4°C overnight. The protein-bound radioactivity was isolated by passing the samples through

Sephadex G-50 columns (1 x 48 cm) in 5 mM potassium phosphate pH 7.3, and was eluted between 10 and 14 ml. Dialysis of sample I against 6 M urea at pH 5.5 reduced the apparently protein-bound radioactivity by 30%; the resulting figure of 4.1 nanomol of bound [^{14}C] carboxamidomethyl group was consistent with a binding of 1 mol of ^{14}C per mol of enzyme, allowing for experimental losses. This bound radioactivity was shown to be present as a glycolamide ester with an enzyme COOH group by the following criteria: (1) exposure to 0.1 M NaOH for 5 h at 20°C caused a 79% loss of bound radioactivity compared with the dialyzed control (1); (2) incubation with 1 M NH_2OH in 6 M urea at pH 9.0 for 5 h at 20°C caused a 55% loss of bound radioactivity (30); (3) incubation with 0.1 M NaOH overnight yielded radioactive material with the mobility of standard glycolic acid on high voltage electrophoresis at pH 2 (1,29).

The bound radioactivity in sample II was diminished by dialysis to a value consistent with the binding of 2 mol of ^{14}C per mol of enzyme. Of this, 50% could be removed by exposure to 0.1 M NaOH, and is evidently present as the glycolamide ester of a COOH group, but the remainder was identified after 6 M HCl hydrolysis as S-[^{14}C] carboxymethylcysteine. Since the assembled evidence was against a role of cysteine, particular care was taken with this identification. Standards of mono- and di-carboxymethylated lysine and histidine and of O-carboxymethyltyrosine and S-carboxymethyl-homocysteine (which results from acid hydrolysis of carboxymethylated methionine residues) were prepared (11-13). On high voltage electrophoresis at pH 2, only O-carboxymethyltyrosine runs closely parallel to S-carboxymethyl-cysteine, and on descending paper chromatography these two are widely separated. The radioactivity released from 0.1 M NaOH-treated-sample II by 6 M HCl hydrolysis for 8 h (7), after purification on a small DEAE Sephadex column with a gradient of 0-1.0 M acetic acid, had a mobility precisely parallel to that of standard S-carboxymethyl-cysteine in both systems.

The results therefore demonstrated that the lysine which can be acetylated by N-acetylimidazole protects a single cysteine residue from reaction with iodoacetamide only when the lysine is the non-acetylated form. If the order of treatments was altered as follows: N-acetylimidazole (to 95% inhibition), acetone precipitation, non-radioactive iodoacetamide plus serine-borate, dialysis, acetone precipitation, [^{14}C] iodoacetamide, then the protein-bound radioactivity corresponded to 2 mol of ^{14}C bound per mol of enzyme; 50% of this radioactivity was labile to 0.1 M NaOH, and 50% was again shown to be S-[^{14}C] carboxymethyl-cysteine. This experiment shows that the cysteine is protected from carboxymethylation also by serine-borate, after the lysine which normally protects it has been acetylated, and therefore suggests that the cysteine is very close to the active site.

A number of reports of covalent binding of 6-diazo-5-oxo-norleucine to the active site of γGT have appeared (15,32). It is possible that this reagent, which normally reacts primarily with SH groups, may be reacting with the shielded cysteine residue.

F. Properties of the γ-Glutamyl-Enzyme Bond

Treatment of γGT with N-acetylimidazole inhibited hydrolysis both of γ-Glu-pNA and of [^{14}C] GSH, but it did not reduce formation of the γ-[^{14}C]-glutamyl-enzyme intermediate (Table 2), suggesting that the acetylated lysine is close enough to interfere with hydrolysis of the intermediate, but not with its formation. Table 2 also shows that serine-borate prevented formation of the γ-glutamyl-enzyme. In a separate control experiment, it was shown that free [^{3}H] glutamic acid did not bind to the fully active enzyme. If the enzyme in the binding experiments shown in Table 2 was saturated, and entirely in the form of the covalent intermediate, as is to be expected with a GSH concentra-

Table 2. Formation of the γ-[^{14}C] glutamyl-enzyme intermediate. Samples of γGT containing 16 units were treated with 25 mM N-acetylimidazole for 1 h, or mixed with 20 mM serine-borate. These samples together with controls were permitted to react for 4 s with 150 μM [^{14}C] Glu-GSH (5 μCi/μmol). The reactions were stopped with 0.5 ml of performic acid and left at 4°C overnight (9). The protein-bound radioactivity was isolated by passing the mixtures through columns of Sephadex G-50 (1 x 50 cm) in 0.1 M sodium acetate, pH 4.0, containing 0.1% (w/v) sodium dodecyl sulfate

Sample	Protein-bound radioactivity (dpm)
Untreated γGT	2,520
γGT + N-acetylimidazole	2,618
γGT + 20 mM serine-borate	nil

tion of 150 μM and a K_m for GSH of about 14 μM (9), then the figures in Table 2 suggest that 1 nanomol of the enzyme corresponds to about 65 γ-Glu-pNA units, in close agreement with the results of Tate and Ross (33).

It may be mentioned that the use of trichloroacetic acid for isolating the labeled enzyme is not effective, since pure γGT is not precipitated by this acid, in common with some other glycoproteins (16). The enzyme is precipitated by acetone (33) but remains active.

Portions of the γ-[^{14}C] glutamyl-enzyme eluted from Sephadex G-50 as in Table 2 were separately treated with 1 M NH_2OH at pH 7.2; 1 M NH_2OH at pH 10; 1 M NH_2OH in 6 M urea at pH 10; 0.5 M NH_3 at pH 10; 0.2 M potassium phosphate, pH 12.0, all for 18 h at 20°C, followed by dialysis against 0.1% sodium dodecyl sulfate. The protein-bound radioactivity was unaffected by all of these treatments except the incubation at pH 12, which released 62% of the bound ^{14}C. The γ-glutamyl-enzyme bond is therefore stable to performic acid, and to NH_2OH and NH_3 up to pH 10. This stability excluded thioester (22,26), acyl-imidazole and phenolic ester (2), and anhydride linkages. It is therefore highly probable that the γ-glutamyl-enzyme intermediate of γGT catalysis contains the isopeptide γ-glutamyl-ε-lysine. This compound has been identified as the basis of the cross-link in fibrin (19,23).

In the case of a classical ping-pong mechanism, the covalent enzyme intermediate can usually be obtained in large amounts simply by omission of the second substrate. With γGT, however, hydrolysis of the γ-glutamyl-enzyme cannot be prevented, though it is much reduced by N-acetylimidazole treatment. This fact, and the enormous cost of [^{14}C] Glu-GSH, have so far limited the amounts of γ-[^{14}C] glutamyl-enzyme which could be isolated to less than 10 nanomol, making further analysis difficult.

G. Conclusion

In summary, the results suggest very strongly that γ-glutamyl-transferase possesses a lysine residue at the active site which binds the γ-glutamyl group derived from the substrate. The enzyme was shown also to possess a COOH group essential to catalytic activity. A second lysine residue was found to be sufficiently close to play some part in hydrolysis of the γ-glutamyl-enzyme intermediate, and to shield a cysteine residue from most SH-directed reagents.

Acknowledgments. This work was partly supported by grants from the Medical Research Council of Canada and from the Faculty of Medicine at Queen's University. The author is grateful to Drs. N.P. Curthoys and T.G. Flynn for helpful discussion and advice.

References

1. Bonner, W.M., Bloch, K.: J. Biol. Chem. 247, 3123-3133 (1972)
2. Bruice, T.C.: Methods Enzymol. 6, 606-610 (1963)
3. Carne, T.J., Flynn, T.G.: Can. J. Biochem. 55, 856-864 (1977)
4. Carraway, K.L., Koshland, D.E.: Biochim. Biophys. Acta 160, 272-274 (1968)
5. Colman, R.F.: J. Biol. Chem. 248, 8137-8143 (1973)
6. Connell, G.E., Adamson, E.D.: Methods Enzymol. 19, 782-789 (1970)
7. Cotner, R.C., Clagett, D.O.: Anal. Biochem. 54, 170-177 (1973)
8. Curthoys, N.P., Kuhlenschmidt, T.: J. Biol. Chem. 250, 2099-2105 (1975)
9. Elce, J.S., Broxmeyer, B.: Biochem. J. 153, 223-232 (1976)
10. Goore, M.Y., Thompson, J.F.: Biochim. Biophys. Acta 132, 15-26 (1967)
11. Gundlach, H.G., Stein, W.H., Moore, S.: J. Biol. Chem. 234, 1754-1760 (1959)
12. Gundlach, H.G., Moore, S., Stein, W.H.: J. Biol. Chem. 234, 1761-1764 (1959)
13. Gurd, F.R.N.: Methods Enzymol. 25, 424-438 (1972)
14. Huc, C., Olomucki, A., Thomé-Beau, F.: FEBS Lett. 60, 414-418 (1975)
15. Inoue, M., Horiuchi, S., Morino, Y.: Eur. J. Biochem. 73, 335-342 (1977)
16. Irwin, D., Anastassiades, T.: Can. J. Biochem. 53, 895-902 (1975)
17. Katunuma, N., Katsunuma, T., Towatori, T., Tomino, I.: In: The enzymes of glutamine metabolism, Prusiner, S., Stadtman, E.R. (eds.) pp. 227-254. New York: Academic Press 1973
18. Klein, S.M., Sagers, R.D.: J. Biol. Chem. 242, 301-305 (1967)
19. Lorand, L.: Ann. N.Y. Acad. Sci. 202, 6-30 (1972)
20. Means, G.E., Feeney, R.E.: Chemical modification of proteins. London and Amsterdam: Holden-Day, Inc. 1971
21. Orlowski, M., Meister, A.: J. Biol. Chem. 240, 338-347 (1965)
22. Perham, R.N., Harris, J.I.: J. Mol. Biol. 7, 316-320 (1963)
23. Pisano, J.J., Bronzert, T.J., Peyton, M.P., Finlayson, J.S.: Ann. N.Y. Acad. Sci. 202, 99-113 (1972)
24. Revel, J.P., Ball, E.G.: J. Biol. Chem. 234, 577-582 (1959)
25. Riordan, J.F., Wacker, W.E.C., Vallee, B.L.: Biochemistry 4, 1758-1765 (1965)
26. Roskoski, R.: Biochemistry 12, 3709-3714 (1973)
27. Stauffer, C.E., Etson, D.: J. Biol. Chem. 244, 5333-5338 (1969)
28. Szasz, G.: Clin. Chem. 15, 124-136 (1969)
29. Szewczuk, A., Connell, G.E.: Biochim. Biophys. Acta 105, 352-367 (1965)
30. Takahashi, K., Stein, W.H., Moore, S.: J. Biol. Chem. 242, 4682-4690 (1967)
31. Takahashi, S., Pollack, J., Seifter, S.: Biochim. Biophys. Acta 371, 71-75 (1974)
32. Takahashi, S., Zukin, R.S., Steinman, H.M.: Federation Proc. 37, Abs. 237 (1978)
33. Tate, S.S., Ross, M.E.: J. Biol. Chem. 252, 6042-6045 (1977)
34. Thompson, J.F.: Meth. Enzymol. 17A, 894-900 (1970)

The Role of Glutathione in Transport Activity in Kidney*

F. H. LEIBACH, D. J. PILLION, J. MENDICINO, and D. PASHLEY

Summary

Diamide oxidizes renal cortical GSH and causes inhibition of amino acid and sugar uptake and gluconeogenesis in the renal cortical slice. However, it does not inhibit the uptake of p-aminohippurate. The effects of diamide can be reversed by the addition of exogenous GSH, while other thiols are less effective. Diamide does not affect renal γ-glutamyl transpeptidase, alkaline phosphatase, or Mg^{2+}-ATPase, but does inhibit protein kinase activity, both soluble and particulate, microsomal Na^{+}-K^{+}-dependent ATPase and glucose-6-phosphatase. The inhibition of these enzymes is partially reversible upon addition of exogenous thiols. The results suggest that GSH is intimately associated with transport processes in kidney.

The suggested functions for glutathione in biological systems are very extensive (1,2,3,4). Glutathione, which is found in high concentration in the kidney, has been directly or indirectly implicated in the maintenance of normal kidney function (see above references). In spite of its generally accepted importance, not enough consideration has always been given to changes in glutathione levels under particular experimental conditions. We have previously reported on the dramatic depletion of GSH during kidney perfusion, and the remarkable efficiency of extraction of GSH from the perfusate, as well as the effect of GSH on certain functional parameters of the isolated perfused rabbit kidney (5,6). Along with the depletion of GSH levels during perfusion there appear to be certain changes in the enzyme profile of the kidney tubule (7), which prompted us to return to the kidney slice for the particular studies we will report at this meeting.

With the introduction of diamide, a new series of investigations into the intracellular function of glutathione was made possible. Diamide was reported to be highly specific for the oxidation of GSH. Addition of diamide to human red blood cells was shown to lead to stoichiometric oxidation of intracellular GSH, and the conversion of GSH → GSSG was very rapid even at low temperature. The effect of diamide on red cell GSH was completely reversible (8).

Our interest in membrane enzymes and in metabolic control processes in the kidney prompted us to explore the effect of diamide on amino acid accumulation in kidney slices (9).

In Table 1 can be seen the effect of diamide on the distribution ratio of a number of amino acids in kidney slices.

It is evident that incubation in the presence of reduced glutathione leads to diverse effects, most notably a 25% increase in the ability of tissues to accumulate [^{14}C] glycine. The effect of diamide treatment

*Supported by USPHS-NIH grants AM 13150 and FR 5365.

Table 1. Effect of diamide on the distribution ratios for several amino acids

Incubation 1	Incubation 2	% of control distribution ratio Arg	Lys	Leu	Gly	Met	α-Amino isobutyric acid
Krebs-Ringer bicarbonate buffer	GSH	113 (4)	95 (15)	97 (8)	125 (4)	112 (8)	101 (14)
Diamide	Krebs-Ringer bicarbonate buffer	67 (10)	59 (9)	55 (8)	63 (4)	59 (9)	54 (15)
Diamide	GSH	87 (6)	87 (11)	83 (8)	94 (4)	83 (8)	73 (14)

Tissues were incubated in Krebs-Ringer bicarbonate buffer with or without diamide ($3 \cdot 10^{-2}$ M) for 15 min at 4°C (Incubation 1). Tissues were then placed in Krebs-Ringer bicarbonate buffer with or without GSH ($2 \cdot 10^{-2}$ M) for 15 min at 25°C (Incubation 2). The tissues were then incubated for 60 min at 37°C in the presence of the labeled and unlabeled amino acid. Samples which were incubated in buffer only served as controls and were assigned a value of 100%. The distribution ratios calculated for the treated samples were then compared with these control values (see (9)). The number of experiments is in parentheses.

however, is shown to be a remarkably consistent reduction of slice to medium distribution ratios for these six amino acids. Kinetic experiments show that diamide affects the initial rate of uptake of amino acids as well as the steady-state level of accumulation. Furthermore, it can be seen that the effect of incubation with diamide is reversible. Thus, diamide-treated slices which are subsequently incubated in the presence of GSH demonstrate a significant recovery in their ability to accumulate amino acids, indicating that the effect of diamide on tissue slices is reversible. Other sulfhydryl compounds are also able to reverse the effect of diamide treatment, but GSH is more effective in this regard than are dithiothreitol or mercaptoethanol. The reversibility of the effects of diamide would strongly indicate that diamide is not lowering amino acid uptake by virtue of tissue toxicity or membrane damage. Additional evidence bearing on this point was obtained by studying p-aminohippuric acid uptake which can be used as an indicator of mitochondrial function (10). Slices were treated in the same manner as shown in Table 1 with or without diamide. Control slices gave a distribution ratio for PAH of 7.51 ± 0.31 whereas diamide-treated slices gave a ratio of 7.13 ± 0.38, demonstrating that diamide-treated slices were capable of concentrative uptake.

In view of these findings with amino acids and the organic acid, PAH, it was of interest to determine whether the effect of diamide affected other transport systems as well (11). For this purpose we chose α-methyl-D-glucoside, a nonmetabolizable sugar which is actively transported by slices of rat kidney cortex and which resembles glucose and galactose in its transport characteristics (12). Table 2 shows the effects of diamide and GSH on the distribution of α-methyl-D-glucoside.

It can be seen that diamide effectively inhibits the uptake of α-methyl-D-glucoside and that GSH can reverse this effect. Furthermore, it was shown that treatment with GSH stimulates the uptake of αMG, whereas the uptake of 3-O-methylglucose, which is not actively accumulated by cortical slices (13) (the distribution ratio remains less than one) is not affected by either treatment. Studies with other thiols indicate that this stimulatory effect of GSH on α-methyl-D-glucoside transport is specific for GSH since mercaptoethanol and dithiothreitol showed no

Table 2. The effect of GSH and diamide on the uptake of α-methyl-D-glucoside and 3-O-methylglucose

Incubation 1	Incubation 2	Distribution ratios	
		α-Methyl-D-glucoside	3-O-methylglucose
Krebs-Ringer bicarbonate buffer	Krebs-Ringer bicarbonate buffer	3.80 ± 0.12 (8)	0.79 ± 0.03 (4)
Krebs-Ringer bicarbonate buffer	GSH	5.24 ± 0.28 (8)	0.71 ± 0.05 (4)
Diamide	Krebs-Ringer bicarbonate buffer	2.09 ± 0.26 (7)	0.79 ± 0.02 (4)
Diamide	GSH	3.66 ± 0.20 (8)	0.82 ± 0.06 (4)

For experimental details see legend to Table 1.

Table 3. The effect of diamide on distribution ratios and protein kinase activity

Substrate	-Diamide	+Diamide	% Inhibition	p values
Arginine	3.13 ± 0.19 (8)	1.98 ± 0.17 (8)	36.7	<0.005
Lysine	3.20 ± 0.29 (6)	1.89 ± 0.03 (6)	41.3	<0.005
Leucine	2.90 ± 0.46 (7)	1.64 ± 0.12 (7)	43.4	<0.0125
Glycine	6.57 ± 0.23 (4)	4.13 ± 0.19 (4)	37.1	<0.005
Methionine	3.38 ± 0.13 (7)	2.17 ± 0.09 (7)	35.8	<0.005
α-Methyl-D-glucoside	3.80 ± 0.12 (8)	2.09 ± 0.26 (7)	38.7	<0.005
Protein kinase activity	16.12 ± 1.02 (5)	9.70 ± 0.94 (5)	39.8	<0.005

For experimental details see legend to Table 1.
Values expressed as pmol ^{32}P incorporated/min.
Protein kinase activity was determined in samples which contained no transport substrate. Values shown represent mean ± s.e.

stimulation; and as in the studies with amino acid uptake, GSH proved the most effective thiol in reversing the inhibition of diamide.

At this point, then, it has been established that (1) GSH stimulates the uptake of α-methyl-D-glucoside and certain amino acids in rat renal cortex, while other thiols do not. (2) Diamide which oxidizes intracellular GSH to GSSG inhibits the uptake of α-methyl-D-glucoside and amino acids, including the nonmetabolizable amino acid α-aminoisobutyric acid, and (3) GSH can reverse this effect of diamide.

The means by which translocation of amino acids and sugars occurs in kidney, and the type of physiological control over the transport process, is still not very well delineated. However, the finding that cyclic nucleotides can stimulate the accumulation of sugars and amino acids (13,14,15) in kidney slices suggested to us the possibility that protein kinases are involved in regulating transport activity.

In order to elucidate a possible relationship between transport and protein kinase activity, rat renal cortical slices were prepared in an identical fashion and assayed for each parameter. The data shown in Table 3 illustrate the direct relationship between the uptake of sugars and amino acids and the level of protein kinase activity. The addition of 30 mM diamide to renal cortical slices leads to a consistent reduction in transport of approximately 40% for a wide variety of amino acids, as well as α-methyl-D-glucoside. It is extremely interesting to note that the addition of the same amount of diamide to the slices leads to a reduction in protein kinase activity which parallels the decreases seen in the distribution ratios. This finding is consistent with the theory that renal protein kinase functions at a rate-limiting step in amino acid and sugar transport. Additional experiments in which this relationship of protein kinase and transport was tested over a wide range of diamide concentrations showed that the extent of inhibition is the same for both parameters at each concentration of diamide. Clearly then there is a direct correlation between the percent inhibition of transport and protein kinase activity at each concentration of diamide (16). Reversibility of the diamide inhibition of protein kinase was also investigated, and it was found that addition of GSH to diamide-treated tissues caused a reversal of the kinase inhibition. Cysteine and mercaptoethanol were equally as effective whereas 2,3-dimercapto-1-propanol was not.

Since cAMP and protein kinase act through protein phosphorylation, we investigated the phosphorylation pattern of kidney membrane preparations (17). By SDS polyacrylamide electrophoresis it was shown that most of the protein-bound phosphate was incorporated into an acidic protein which migrated toward the end of the gel. Recent experiments have shown that the phosphorylation of this acidic protein specifically is inhibited by diamide and that the inhibition is reversible by GSH and other thiols. Further studies on this peptide and its possible relationship to transport phenomena are currently being done.

Since all our results indicate that diamide is acting at the membrane, we decided to examine its effects on enzymes present in different membrane fractions prepared from kidney cortex.

Table 4 shows that bound protein kinases present in brush border and microsome membrane fractions were inhibited by diamide. Only a slight inhibition was observed at 1.5 mM diamide whereas about 50% inhibition was obtained with 4.5 mM diamide. Glucose-6-phosphatase and Na^{+}-K^{+} dependent ATPase were also inhibited. Other enzymes present in these particulate fractions were not inhibited by diamide. Thus, the activities of alkaline phosphatase, Mg^{2+}-dependent ATPase and γ-glutamyl transpeptidase were not altered by the addition of diamide. Clearly, the inhibition shows some specificity for enzymes present in these membrane fractions. It is not certain whether diamide is acting directly on glucose-6-phosphatase and Na^{+}-K^{+} ATPase or whether the observed decreases are indirectly related to the inhibition of protein kinases in these preparations. It should be noted that the extent of inhibition of these enzymes at different concentrations of diamide were similar to those observed for bound protein kinase. The inhibition of glucose-6-phosphatase led us to an investigation of the effects of diamide on gluconeogenesis. A role in glucose transport has been suggested for this enzyme in liver and kidney (18). In Table 5 it can be seen that renal gluconeogenesis from pyruvate was indeed inhibited by diamide, and that the inhibition can be reversed by GSH.

Since it is well established that Na^{+} is cotransported with amino acids and sugars (19) the finding that diamide inhibits the Na^{+}-K^{+} ATPase becomes significant and we thus investigated this point more fully. Previous reports in the literature had indicated that GSH-oxidizing agents do affect cation gradients and possibly ATPase activity. In the isolated hemoglobin-free perfused rat liver a reversible release of K^{+} upon the addition of t-butyl hydroperoxide was demonstrated (20) and more recently it was demonstrated that GSH oxidation by tertiary butyl

Table 4. Effect of diamide on the activity of bound enzymes present in renal brush border and microsomal membrane fractions

Enzyme	Diamide present	Subcellular fraction	
		Brush border	Microsomes
	mM	pmol · min^{-1} · mg $protein^{-1}$	
Protein kinase	none	18.2 ± 3.1	15.4 ± 2.3
	1.5	17.5 ± 2.3	11.9 ± 0.7
	4.5	9.1 ± 1.3	7.3 ± 0.9
		µmol · min^{-1} · mg $protein^{-1}$	
Glucose-6-phosphatase	none	-	8.2 ± 0.2
	1	-	7.8 ± 0.5
	5	-	4.1 ± 0.3
Alkaline phosphatase	none	4.5 ± 0.3	14.3 ± 1.0
	1	4.7 ± 0.5	16.2 ± 1.8
	8	4.3 ± 0.2	13.7 ± 0.9
Na^+,K^+-ATPase	none	-	35.5 ± 2.9
	0.5	-	30.1 ± 2.4
	3.0	-	20.7 ± 4.3
Mg^{2+}-ATPase	none	6.8 ± 0.9	157.8 ±12.7
	0.5	6.7 ± 0.7	150.2 ±11.0
	3.0	6.4 ± 0.3	149.7 ±12.7
γ-Glutamyl-transpeptidase	none	-	43.1 ± 3.7
	0.5	-	41.0 ± 3.6
	5.0	-	44.2 ± 4.3

The subcellular fractions were prepared from fresh rat kidneys as described in the text. When indicated, diamide was added to the incubation mixture before initiating the reaction with substrate. Results are expressed as µmol of product formed min^{-1}, except those for protein kinase which are expressed as pmol·min^{-1}·mg $protein^{-1}$.

Table 5. The effects of diamide and GSH on gluconeogenesis from pyruvate

Incubation I	Incubation II	Gluconeogenesis µmol/gm/90 min	Protein kinase activity (pmol/min)
Buffer	Buffer	52.88 ± 1.7	33.04 ± 1.5
Diamide	Buffer	6.38 ± 2.2	18.12 ± 1.5
Buffer	GSH	60.66 ± 6.6	31.97 ± 0.9
Diamide	GSH	47.27 ± 1.6	26.36 ± 0.8

For experimental details see legend to Table 1.
Pyruvate concentration was 10 mM. Glucose was determined by the glucose oxidase method.

hydroperoxide also causes a decrease in the Na^+-K^+ ATPase activity of lens epithelium (21).

In Table 6 are shown the effects of diamide on kidney slice Na^+ content and Na^+-K^+-dependent ATPase. Slices were prepared and incubated as in transport studies. Treatment with increasing diamide concentrations causes first a decrease and then a significant increase in tissue Na^+ level. This latter result would be expected if diamide were capable of inhibiting Na^+-K^+ ATPase activity.

Table 6. Effect of diamide on tissue sodium content and Na^+-K^+ ATPase activity

Diamide [mM]	Na^+ (mEq/kg dry wt)	Na^+-K^+-ATPase (µg Pi/mg protein/5 min)
0	285.1 ± 30.0	13.7 ± 0.4
5	222.5 ± 16.2	13.7 ± 0.8
10	183.4 ± 19.0[a]	13.8 ± 0.9
30	382.9 ± 44.8	7.6 ± 2.4[a]
50	564.3 ± 23.9[a]	7.4 ± 2.2[a]
70	712.8 ± 186.4[a]	7.3 ± 0.1[a]

[a] $P < 0.05$.

Slices were incubated with various concentrations of diamide (Incubation I) and then incubated in buffer (Incubation 2 and uptake period) containing ^{14}C-inulin. At the end of the 60 min uptake period, slices were removed from the buffer and paired samples were used to determine intracellular sodium levels (i.e. corrected for inulin space) and Na^+-K^+-ATPase activity. Results are expressed as means ± SEM for triplicate determinations.

Table 7. The effect of diamide on reduced glutathione and total glutathione levels in rat kidney cortex slices

	- Diamide			+ Diamide		
Time	Total Glutathione[a]	GSH	GSSG[b]	Total Glutathione	GSH	GSSG[b]
-60	1.02	0.88	0.14	c	c	c
-30	0.81	0.72	0.09	c	c	c
-15	c	c	c	0.47	0.03	0.43
0	0.72	0.63	0.09	0.36	0.37	0.00
30	0.56	0.51	0.05	0.37	0.40	0.00
60	0.50	0.37	0.13	0.33	0.28	0.05

For experimental details see legend to Table 1.
Time 0 refers to start of incubation at 37°C, while time -60 refers to fresh tissue.
[a] All data expressed as mg glutathione/g wet weight of tissue.
[b] Calculated as the difference between total glutathione and GSH.
[c] Not determined.

To test the effects of diamide on the levels of kidney cortex GSH and GSSG, we incubated slices in a manner identical to that used in the uptake experiments.

The data in Table 7 indicate that GSSG levels in untreated cortical slices are usually much lower than GSH levels, although GSH levels decline and GSSG levels rise after a 60-min incubation at 37°C. In diamide-treated slices, GSSG levels are elevated while diamide is present, but after removal of the diamide, the GSSG levels are extremely low. When one compares total glutathione, after a 60-min incubation it is lower in the diamide-treated tissue and the possibility has to be considered that the GSH which disappears from the diamide-treated slice is tied up in the form of mixed disulfides with protein, as suggested by previous studies (22,23). This is all the more likely since analysis of the incubation media did not show any GSSG present.

In conclusion, a lowering of the GSH level in slices of kidney cortex leads to the reversible inhibition of protein kinases, Na^+-K^+-dependent

ATPase (24), glucose-6-phosphatase, amino acid and α-methyl-D-glucoside uptake and gluconeogenesis. The exact role of these enzymes in the regulation of transport and gluconeogenesis is not presently known, however, the results obtained thus far suggest that rate-limiting steps involved in the transfer of substrates in membranes may be the site of action of protein kinases and that GSH is required for expression of full activity of the transport systems.

References

1. Colowick, S., Lazarow, A., Racker, E., Schwartz, D.R., Stadtman, E., Waelsh, H. (eds.): Glutathione. New York: Academic Press 1954
2. Crook, E.M. (ed.): Glutathione. London: Cambridge University Press 1959
3. Flohé, L., Benöhr, H.C., Sies, H., Waller, H.D., Wendel, A. (eds.): Glutathione. New York: Academic Press 1974
4. Arias, I.M., Jakoby, W.B. (eds.): Glutathione. New York: Raven Press 1976
5. Leibach, F.H., Fonteles, M.C., Pillion, D.J., Karow, A.M.: Glutathione in the isolated perfused rabbit kidney. J. Surg. Res. 17, 228-231 (1974)
6. Fonteles, M.C., Pillion, D.J., Jeske, A.H., Leibach, F.H.: Extraction of glutathione by the isolated perfused rabbit kidney. J. Surg. Res. 21, 169-174 (1976)
7. Pillion, D.J., Jeske, A.H., Leibach, F.H.: γ-Glutamyl transpeptidase in the urine from an isolated rabbit kidney perfused with and without DMSO. Biochem. Pharmacol. 25, 913-918 (1976)
8. Kosower, N.S., Kosower, E.M., Wertheim, B., Correa, W.S.: Diamide, a new reagent for the intracellular oxidation of glutathione to the disulfide. Biochem. Biophys. Res. Commun. 37, 593-596 (1969)
9. Hewitt, J., Pillion, D., Leibach, F.H.: Inhibition of amino acid accumulation in slices of rat kidney cortex by diamide. Biochim. Biophys. Acta 363, 267-276 (1974)
10. Reimer, K.A., Jennings, R.B.: Alterations in renal cortex following ischemic injury. PAH uptake by slices of cortex after ischemia or autolysis. Lab. Invest. 25, 176-184 (1971)
11. Pillion, D.J., Leibach, F.H.: The effect of diamide and glutathione on the uptake of α-methyl-D-glucoside by slices of rat kidney cortex. Biochim. Biophys. Acta 382, 246-252 (1975)
12. Segal, S., Rosenhagen, M., Rea, C.: Developmental and other characteristics of α-methyl-D-glucoside transport by rat kidney cortex slices. Biochim. Biophys. Acta 291, 519-530 (1973)
13. Rea, C., Segal, S.: Dibutyryl cyclic adenosine monophosphate enhancement of α-methyl-D-glucoside accumulation by kidney cortex slices. Biochim. Biophys. Acta 311, 615-624 (1973)
14. Weis, I.W., Morgan, K., Phang, J.M.: Cyclic adenosine monophosphate stimulates transport of amino acids in kidney cortex. J. Biol. Chem. 247, 760-764 (1972)
15. Reynolds, R., Segal, S.: Effects of dibutyryl cyclic AMP on α-methyl-D-glucoside accumulation in rabbit kidney. Am. J. Physiol. 226, 791-795 (1974)
16. Pillion, D.J., Leibach, F.H., von Tersch, J., Mendicino, J.: Inhibition of protein kinase activity and amino acid and α-methyl-D-glucoside transport by diamide. Biochim. Biophys. Acta 419, 104-111 (1976)
17. Abou-Issa, H., Mendicino, J., Pillion, D.J., Leibach, F.H.: Phosphorylation and dephosphorylation of renal brush border membranes by protein kinase and phosphoprotein phosphatase. FEBS Lett. 50, 121-124 (1975)
18. Nordlie, R.: Metabolic regulation by multifunctional glucose-6-phosphatase, Curr. Top. Cell. Regul. 8, 33-117 (1974)
19. Schultz, S.G., Curran, P.F.: Coupled transport of sodium and organic solutes. Physiol. Rev. 50, 637-718 (1970)
20. Sies, H., Gerstenecker, C., Summer, K.H., Menzel, H., Flohé, L.: Glutathione-dependent hydroperoxide metabolism and associated metabolic transitions in hemoglobin-free perfused rat liver. In: Glutathione. Flohé, L. et al. (eds.) pp. 261-275. New York: Academic Press 1974

21. Giblin, F.J., Chakrapani, B., Reddy, V.N.: Glutathione and lens epithelial function. Invest. Ophthamol. 15, 381-393 (1976)
22. Harrap, K.R., Jackson, R.C., Riches, P.G., Smith, C.A., Hill, B.T.: The occurrence of protein-bound mixed disulfides in rat tissues. Biochim. Biophys. Acta 310, 104-110 (1973)
23. Mannervik, B., Axelsson, K.: Reduction of disulphide bonds in proteins and protein mixed disulphides catalysed by a thioltransferase in rat liver cytosol. Biochem. J. 149, 785-788 (1975)
24. Pillion, D.J., Moree, L., Rocha, H., Pashley, D., Mendicino, J., Leibach, F.H.: The role of glutathione in renal cortical tissue. Effects of diamide on Na^+ and GSSG levels, amino acid transport and Na^+-K^+-ATPase activity. Mol. and Cell. Biochem. 18, 109-115 (1977)

Chapter III
Hydroperoxide and Disulfide Metabolism

Hydroperoxide Metabolism: An Overview

B. CHANCE, A. BOVERIS, Y. NAKASE, and H. SIES

Summary

The well-documented generation of the oxygen intermediates O_2^- and H_2O_2 at a multiplicity of cellular sources in biological systems may lead to the production of the unstable and reactive oxygen species HO• and $^1O_2$. These latter species are able to initiate a free radical chain reaction that leads to lipid and organic peroxide formation. The response of rat liver to hyperbaric oxygen, for which most of the physiological and biochemical parameters related to hydroperoxide metabolism have been verified, supports this interpretation. Increased oxygen tensions produce (1) increased O_2^- generation by the mitochondria and cytosolic enzymes; (2) increased H_2O_2 production in isolated mitochondria, microsomes and peroxisomes, and in isolated rat liver cells; (3) a very slight increase of H_2O_2 production in perfused liver and in the organ in situ, indicating in the latter case the protective role of the microvascular system; and (4) an increase in GSSG release from the perfused rat liver, which is especially marked in the tocopherol-deficient liver. Oxidized glutathione release, reflecting the presence of intracellular glutathione peroxidase substrates, affords a very sensitive assay for oxidative stress. Glutathione peroxidase turnover is linked to the main redox state of the cell by the NADPH-dependent glutathione reductase and the transhydrogenases. Chemiluminescence from perfused rat liver and from the isolated subcellular fractions is markedly stimulated by lipid hydroperoxides. Studies of chemiluminescence and ethane formation may prove useful approaches for understanding lipid peroxidation in intact systems.

A. The Primary Metabolism of Oxygen Intermediates

It is now well established that, although present in low concentrations, hydrogen peroxide and superoxide anion are normal metabolites in the aerobic cell. A remarkable multiplicity of cellular sources is found (Fig. 1). Generation of H_2O_2 by product-specific oxidases seems to provide the teleological reason for their association with catalase in the peroxisomes (1-3). These organelles can be considered to be the chief source of H_2O_2 in rat liver. However, only a minor fraction of peroxisomal H_2O_2 is able to diffuse out to the cytosol, most of it being utilized by intraorganellular catalase (4-6). Mitochondria (7-9) and the membranes of the endoplasmic reticulum (8-11) are apparently the most important sources of cytosolic H_2O_2 to maintain the steady state level. Soluble cytosolic enzymes appear as minor contributors to total H_2O_2 production in rat liver (6,9). The panoply of cellular sources for H_2O_2 is modulated by a series of controls, both physiological and biochemical, such as the transition from the resting State 4 to the active State 3 and the supply of oxygen and substrate.

Superoxide anion is generated in the liver cell by flavin enzymes and iron-sulfur proteins in the cytosol (12-15) and by the auto-oxidation of reduced ubiquinone in the mitochondria (16-17). The rates of H_2O_2

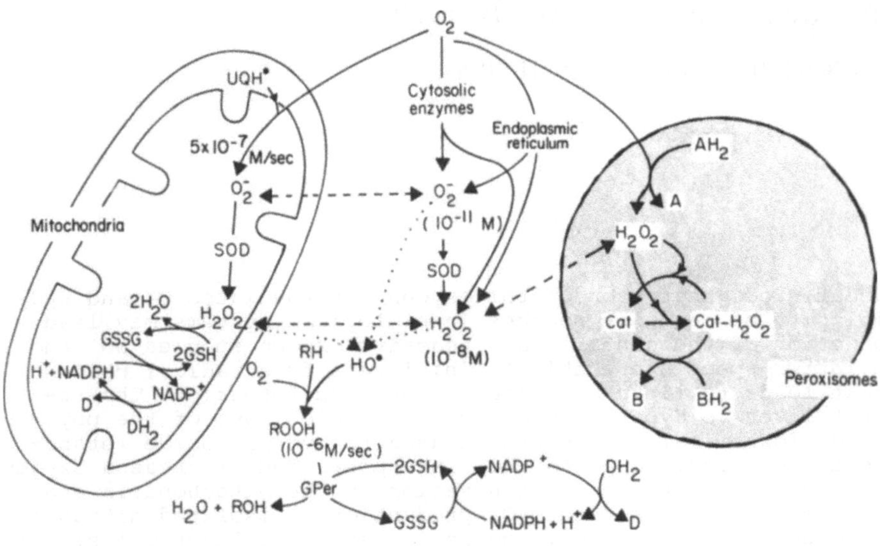

Fig. 1. A general diagram indicating the sources and sinks for oxygen reduction products in the mitochondrial, cytosolic, and peroxisomal spaces. Abbreviations are: *UQH*•, ubiquinone radical; *GSSG*, oxidized glutathione; *GSH*, reduced glutathione; DH_2 and *D*, unspecified NADP-reducing systems; *SOD*, superoxide dismutase; *NADPH* and *NADP*, reduced and oxidized nicotinamide adenine dinucleotide phosphate; O_2^-, superoxide anion; *HO*•, hydroxyl radical; *ROOH*, an alkyl hydroperoxide; *GPer*, glutathione peroxidase; *Cat*, catalase; *B* and BH_2, hydrogen donors of a specificity appropriate to catalase, such as ethanol

generation in the subcellular fractions that include primary O_2^- formation (9) suggest that the multi-enzyme redox systems of mitochondrial and endoplasmic reticulum membranes are the chief sources of O_2^- in the rat liver cell, and that cytosolic enzymes such as xanthine oxidase and aldehyde oxidase are less effective. The level of superoxide anion, the more reactive species, is maintained at 10^{-12} to 10^{-11} M by superoxide dismutase (18), while that of hydrogen peroxide, which is less reactive, is regulated by catalase and glutathione peroxidase at concentrations up to three orders of magnitude greater, 10^{-9} to 10^{-7} M, depending on H_2O_2 production (19). These protective enzymes are present in aerobic tissues at high levels, their concentrations being higher than those of their substrates; they act accordingly as scavenger enzymes. For instance, catalase and superoxide dismutase levels in rat liver have been estimated at about 10^{-5} M in the whole tissue (18,19), a value that exceeds substrate concentration by about three and six orders of magnitude, respectively.

There has been much discussion as to whether catalase or glutathione peroxidase is the predominant enzyme in regulating intracellular H_2O_2 levels. Both fulfill important metabolic functions in controlling H_2O_2 concentrations at different levels and in different parts of the cell (Fig. 1). Catalase is especially effective as a "safety valve" for dealing with the large amounts of H_2O_2 that may be generated in the peroxisomes. Glutathione peroxidase is capable not only of utilizing organic hydroperoxides but also of metabolizing H_2O_2 in both the cytosolic and mitochondrial compartments. In fact, no enzymatic pathway is used completely to the exclusion of the other in metabolizing H_2O_2. In rat liver, with a GSH concentration of about 8 mM, the relative reaction rates for catalase and glutathione peroxidase depend upon the localized concentrations of both enzymes and of H_2O_2 within the cell. The compartmentation of catalase in the peroxisomes and of glutathione peroxidase in the cytosol and the mitochondria facilitates their effective collaboration in H_2O_2 metabolism, each enzyme being chiefly responsible for the decomposition of H_2O_2 generated at the intracellular site at which the enzyme is located. However, it has been shown that addition of peroxisomal substrate, glycolate, to perfused liver activates glutathione peroxidase (6,20). This response could indicate the presence of cytosolic glycolate oxidase, or, more probably, the existence of at least minimal intracellular H_2O_2 gradients.

B. Lipid and Organic Peroxide Formation and Glutathione Release

Although the very existence of lipid and organic peroxide formation under in vivo conditions is not universally accepted, and although the rates of in vivo lipoperoxidation are under debate, recent data provided by a series of laboratories in quite independent fields fill the gaps of earlier work. There is at present ample evidence that the intermediates of oxygen reduction, especially superoxide anion (which in itself constitutes a source of hydrogen peroxide), are able to initiate a chain reaction that leads to lipoperoxide formation (21-26). The hypothesis that the interaction of H_2O_2 and O_2^- in vivo leads to the formation of hydroxyl radical ($HO^\bullet$) and singlet oxygen ($^1O_2$), which is the rate-limiting step for lipoperoxidation, seems to fit most of the data. The protective effects of superoxide dismutase (22-26), catalase (22-26), $HO^\bullet$ scavengers (22,24) and $^1O_2$ quenchers (22,23) thus appear reasonably explained.

Sies et al. (27-29) have found that addition of hydroperoxide to isolated liver cells and to perfused liver leads to an increased release of oxidized glutathione, *t*-butyl and cumene hydroperoxide being especially effective (Fig. 2). Measurement of oxidized glutathione release provided a major breakthrough for the estimation of glutathione peroxidase turnover, and therefore of formation of glutathione peroxidase substrate, a process that collects the products of cellular over-oxidations. This approach has provided valuable data for an understanding of the mechanism of lipoperoxide formation in biological systems. Perfused rat liver releases 1 to 2 nmol GSSG/min per g of liver (6,28,30), a value corresponding to an endogenous lipid peroxide production of approximately 30 to 50 nmol/min per g of liver, a calculation made on the basis that GSSG release reflects about 3% of the turnover of glutathione peroxidase (29).

C. Defense Mechanisms

There are a series of physiological and biochemical mechanisms that act to oppose a deleterious accumulation of oxygen intermediates and hydroperoxides. Tissue oxygen tension is maintained at rather low levels (1-10 torr) primarily by the sensitivity of the microvascular system, which either reduces or increases blood flow in response to either increased or decreased oxygen tensions. The steep oxygen gradient in the tissue (31,32) also contributes to maintaining a low oxygen tension in most of the cells. Another special protection is afforded by the molecular design of cytochrome oxidase, which retains the intermediates of oxygen reduction until water is formed (33). Superoxide dismutase,

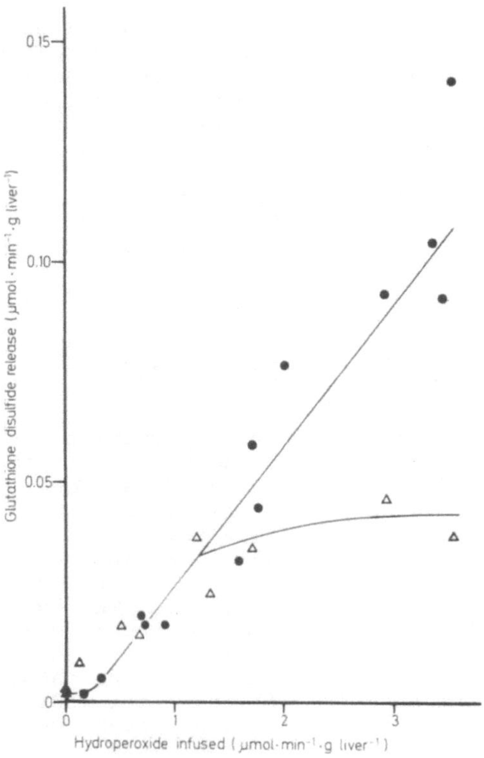

Fig. 2. Dependence of the rate of GSSG release upon the rate of hydroperoxide infusion in perfused rat liver. Maximal GSSG concentrations, obtained about 3 to 5 min after onset of hydroperoxide infusion, were plotted. *t*-Butyl hydroperoxide (●) and H_2O_2 (Δ) data are shown

catalase, and glutathione peroxidase action is directed toward maintaining low intracellular levels of O_2^- and H_2O_2 (Fig. 1), thereby minimizing the rate of the free radical chain reaction that leads to lipoperoxide formation.

Liposoluble molecules such as α-tocopherol (34) and β-carotene (35) could act as $HO^{\bullet}$ scavengers and $^1O_2$ quenchers in sites where lipoperoxidation is likely to occur. In addition, glutathione peroxidase is able to detoxify lipoperoxide to hydroxyfatty acid derivatives (Fig. 1).

D. The Hyperoxic Oxidative Stress

Hyperoxic stress is one experimental condition in which most of the physiological and biochemical parameters related to hydroperoxide metabolism have been verified. Increased oxygen tensions are toxic and lethal to all biological species (36); such universal effect must be related to a very basic metabolic response (37). The rat liver, as a model, has been studied in detail; it has a high content of superoxide dismutase, catalase, and glutathione peroxidase, and thus should have the maximal protection that a mammalian organ can afford against oxygen toxicity.

Mitochondrial production of O_2^- depends almost linearly upon oxygen tension (Fig. 3). Metallo-flavoproteins also shown enhanced O_2^- production under increased oxygen concentrations (38). Thus, increased O_2^-

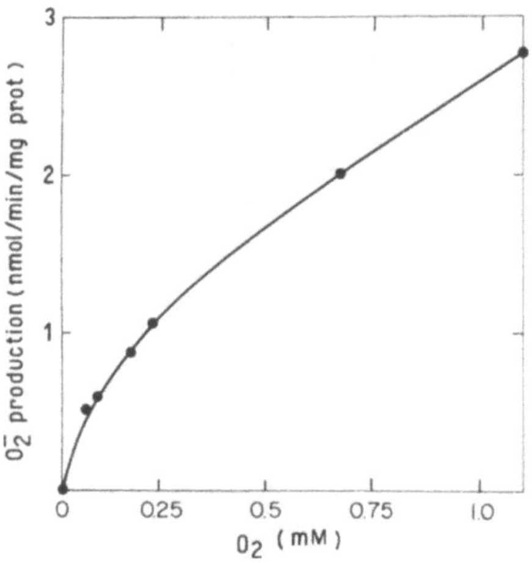

Fig. 3. Oxygen dependence of O_2^- production in beef heart submitochondrial particles. The mannitol-sucrose-50 mM Tris-Mops buffer, pH 7.4, containing the particles and 1 mM epinephrine was equilibrated with oxygen-nitrogen gas mixtures of known composition

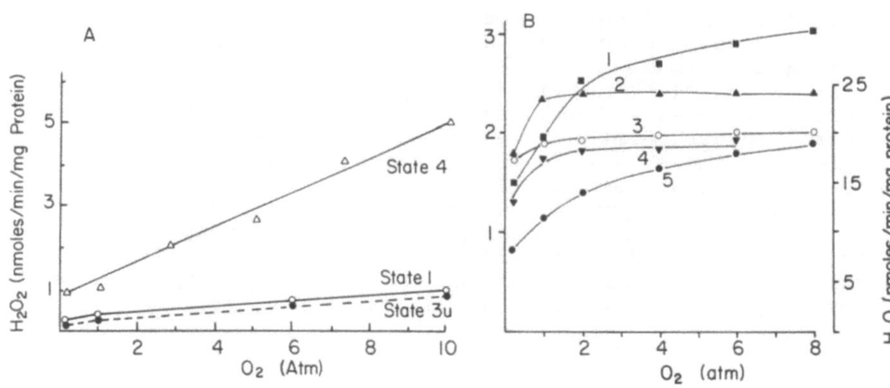

Fig. 4A and B. Effect of hyperbaric oxygen on the production of H_2O_2 by rat liver mitochondria (A) and rat liver peroxisomes and microsomes (B). *1*, Intact peroxisomes with endogenous substrate; *2*, deoxycholate-treated peroxisomes + 20 µM uric acid; *3*, microsomes + 40 µM NADPH; *4*, deoxycholate-treated peroxisomes + 5 mM D-alanine; *5*, deoxycholate-treated peroxisomes + 2 mM glycolate. *Left ordinate* for lines *1*, *3* and *4*; *right ordinate* for lines *2* and *5*

production will immediately follow increased oxygen tension. Hydrogen peroxide production by isolated rat liver organelles has been determined under hyperoxic and hyperbaric oxygen conditions. Mitochondrial production of H_2O_2 is markedly increased in this oxidative stress (Fig. 4), whereas other subcellular structures such as peroxisomes and the endoplasmic reticulum show a limited enhancement in the hyperbaric regions (Fig. 4). The isolated rat liver cells show two- and three-fold increases in H_2O_2 production under hyperbaric oxygen [Fig. 5; (37)], in agreement with the response of the isolated organelles.

However, at the next level of biological organization, in the perfused liver, there is a remarkable lack of marked response in the H_2O_2 production under hyperbaric oxygen (5,39). Limitation of peroxisomal substrates and lack of both extrahepatic metabolites and hormonal stimuli

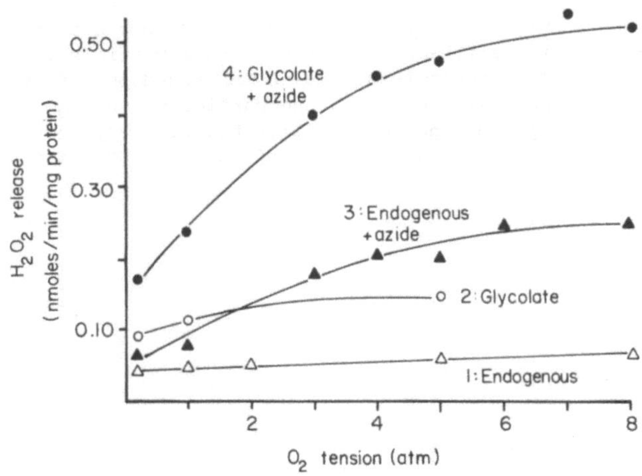

Fig. 5. Effect of hyperbaric oxygen on hydrogen peroxide release from isolated rat liver cells, measured by H_2O_2 binding to yeast cytochrome c peroxidase

could partially explain the lack of stimulation; however, it is reasonable to assume that the catalase reaction in the peroxisomes is not in equilibrium with the cellular H_2O_2 (6). Oshino et al. (40) extended their studies to the rat liver in situ, showing that hyperbaric oxygen only slightly increases (about 10%) H_2O_2 production, and thus confirmed the lack of marked response in the presence of an active microvascular system. Nishiki et al. (39) measured H_2O_2 production, as detected by the peroxisomal catalase, and GSSG release in the perfused rat liver under hyperbaric oxygen. Glutathione release was far more affected than H_2O_2 production and showed a substantial stimulation of hydroperoxide metabolism under hyperoxic stress (Fig. 6). These biochemical studies, that include normal and tocopherol-deficient animals, further identified vitamin E as an important nutritional protective factor against oxygen toxicity. Tocopherol-deficient rat livers show a 2.5 times increased H_2O_2 production and a 3.5 times enhanced GSSG release upon exposure to hyperbaric oxygen, as compared with normal rat livers (39).

E. Hydroperoxide Reduction and Pyridine Nucleotide Oxidation

As shown in Figure 7, the addition of a small amount of an organic hydroperoxide such as *t*-butyl hydroperoxide to perfused liver or isolated hepatocytes leads to a rapid and substantial temporary decrease of pyridine nucleotide absorption due to oxidation of NADPH (27-29), with a subsequent recovery. This sequence would be expected from the scheme of Figure 1. Hydroperoxide reduction by the action of GSH and glutathione peroxidase is coupled to NADPH oxidation by the activity of the NADPH-dependent glutathione reductase. NADPH is coupled to the mitochondrial redox state by the transhydrogenase pathway. It can be seen in Figure 8 that pyridine nucleotide oxidation accompanies GSH oxidation and is proportional to the amount of added hydroperoxide (6,27-29). Increased GSSG release occurs simultaneously with pyridine nucleotide oxidation upon addition of hydroperoxides (*t*-butyl hydroperoxide or H_2O_2) to the perfused rat liver. The extent of both responses is much more marked upon infusion of *t*-butyl hydroperoxide than upon addition of H_2O_2, due to the partial decomposition of H_2O_2 by catalase (6). With high concentrations of added hydroperoxide, NADPH is oxidized to such an extent that the remaining concentration of NADPH falls to the minimal steady-state level observed in intact cells; at this level secondary effects,

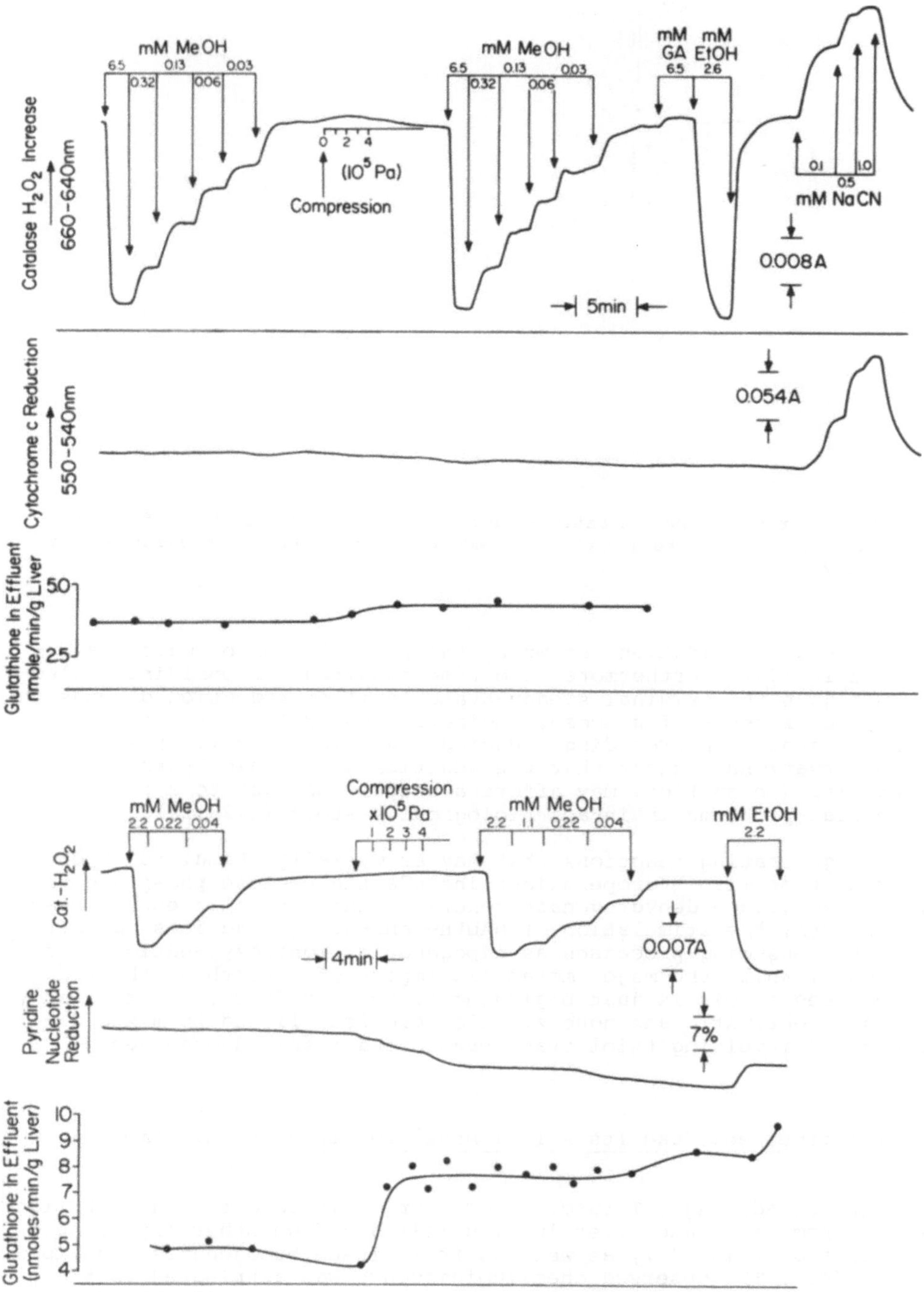

Fig. 6. Hydrogen peroxide production and oxidized glutathione release under normoxic and hyperoxic conditions in the perfused liver from normal (*top graph*) and tocopherol-deficient (*lower graph*) rats. Experimental conditions are detailed in (39)

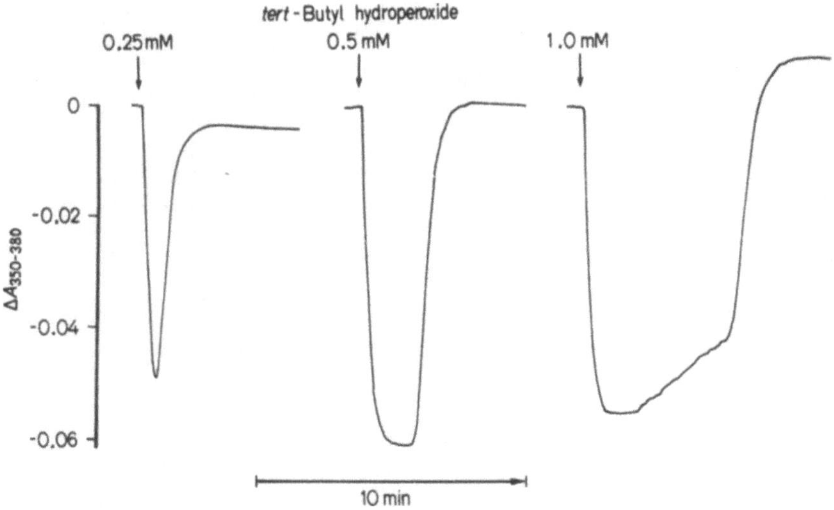

Fig. 7. Time course of NADPH oxidation-reduction changes upon addition of *t*-butyl hydroperoxide to isolated hepatocytes, as monitored by absorbance difference spectrophotometry

such as lipid peroxidation (shown by the accumulation of malondialdehyde) set in (29). Furthermore, the time required for pyridine nucleotide to regain the original steady-state level of reduction depends in part on the presence of glucose, indicating the role of the pentose phosphate pathway in providing reducing equivalents (29). These and other observations suggest that the addition of organic hydroperoxides, applied with due caution, may afford a useful approach to a number of experimental problems in intact biological systems (6,28,29).

The NADPH-generating reactions that may be markedly stimulated upon supplementation with hydroperoxides include the pentose phosphate pathway, the isocitrate dehydrogenase reaction, and the malic enzyme reaction. In turn, the stimulation of NADPH-generating reactions may affect such NADPH-consuming processes as lipogenesis, mono-oxygenations and ureogenesis. While the major metabolic impact of perturbing the redox state of free thiols is just beginning to be elucidated, it is clearly mediated by enzymatic and nonenzymatic reactions linked to mixed disulfides and involving thiol transferases and transhydrogenases.

F. Chemiluminescence and Its Relationship to Lipid Peroxidation

Almost two decades ago, Tarusov et al. were able to detect a weak light emission from the mouse liver in situ (41) and from other organs such as muscle and brain (42), as well as from tissue homogenates and lipid extracts (43). The observed chemiluminescence was attributed to the oxidative chain reactions involving biological lipids (42,43). Some years after, Howes and Steele (44,45), and more recently Nakano et al. (46,47), reported on the chemiluminescence of isolated rat liver microsomes. Both groups related their observations to the level of lipid peroxide, and for optimal photoemission they utilized the conditions that produce microsomal lipoperoxidation (48).

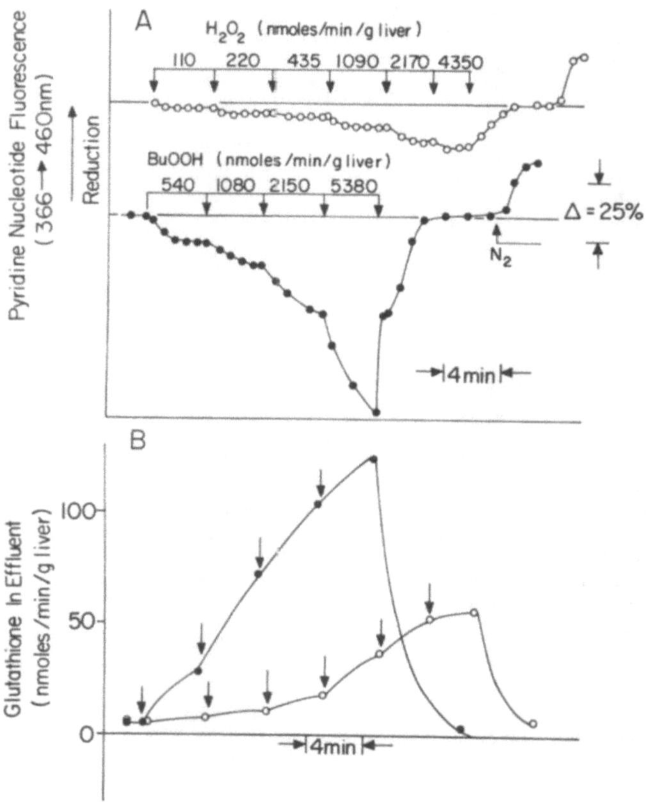

Fig. 8A and B. Effect of H_2O_2 and *t*-butyl hydroperoxide (A) on the oxidation-reduction state of nicotinamide adenine dinucleotide and (B) on the rate of GSSG release from the perfused liver. H_2O_2 (o) and *t*-butyl hydroperoxide (•) were infused at the rates indicated in the figure. The oxidized glutathione concentration in the effluent is expressed in terms of GSH equivalents in (B)

Figure 9 shows that addition of *t*-butyl hydroperoxide to the perfusion fluid produced a detectable photoemission from the perfused rat liver. Perfusion of substances able to increase or decrease intracellular levels of H_2O_2, such as urate and alcohol, did not produce significant changes in liver chemiluminescence. It was found that photoemission was oxygen-dependent; anoxia suppressed the chemiluminescence induced by *t*-butyl hydroperoxide.

Rat liver mitochondria and microsomes supplemented with their physiological substrates also showed a weak photoemission (Table 1). Concerning mitochondria, chemiluminescence was higher in the aerobic reduced states, such as State 4, or in the presence of antimycin, as compared with the more oxidized States 1 and 3u, and was absent in the anoxic State 5. Supplementation with *t*-butyl hydroperoxide in the presence of oxygen increased mitochondrial chemiluminescence, the more marked effect being observed in the State 3u mitochondria, whereas no stimulation occurred in the anoxic condition.

Microsomal chemiluminescence was detectable after addition of NADPH as substrate. Again with the microsome, *t*-butyl hydroperoxide was able to increase the luminescent signal markedly. Microsomal photoemission is also enhanced by supplementation with Fe^{2+} ions and ADP [Table 1; (46)]. The homogenate and the supernatant are almost inactive, and they do not show chemiluminescence upon addition of *t*-butyl hydroperoxide.

The apparent requirements for mitochondrial and microsomal chemiluminescence are: (1) a hydroperoxide; (2) molecular oxygen; and (3) an

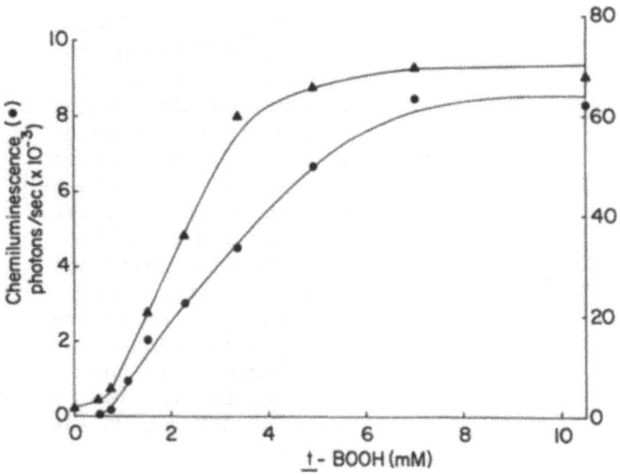

Fig. 9. Chemiluminescence of the perfused rat liver upon infusion of *t*-butyl hydroperoxide. The ordinate corresponds to photons/s over the counts measured before adding the hydroperoxide; *t*-butyl hydroperoxide concentration in the perfusion fluid is indicated in the abscissa. An RCA 8850 photomultiplier was used, cooled down to -30°C. The output was fed into an amplifier and discriminator adjusted for single photon counting. The perfused liver was held at 50 mm from the photomultiplier aperture. Thermal isolation was achieved with a twopane Plexiglas window

Table 1. Chemiluminescence from subcellular fractions of rat liver

	Photons/s	
	Starting conditions	+4 mM *t*-BOOH
Mitochondria		
Background (buffer)	18 ± 1	18 ± 1
State 1	24 ± 2	73 ± 4
State 4 (10 mM succinate)	29 ± 2	83 ± 4
idem + 2 μM antimycin	26 ± 1	68 ± 6
State 3u (10 mM succinate + 2 μM FCCP)	22 ± 2	137 ± 6
State 5	20 ± 1	21 ± 1
same + 1 mM H_2O_2	27 ± 1	158 ± 10
Microsomes		
Background (buffer)	16 ± 1	16 ± 1
Microsomes	21 ± 2	135 ± 4
+ 0.24 mM NADPH	32 ± 2	148 ± 5
+ NADPH + 80 μM $FeSO_4$ + 1 mM ADP	138 ± 5	180 ± 3
Homogenate and supernatant		
Background (cuvette)	14 ± 1	14 ± 1
Homogenate	16 ± 3	15 ± 2
Supernatant	16 ± 2	15 ± 2

Mitochondria (1.9 mg of protein/ml) were suspended in 0.23 M mannitol, 0.07 M sucrose, 25 mM Mops-KOH, pH 7.3. Microsomes (1.4 mg of protein/ml) were suspended in 90 mM K_2HPO_4-NaH_2PO_4 buffer, pH 7.3. Homogenate and 105,000 *g* supernatant were used without additions. Final volume was 5.0 ml. Air was bubbled through the samples to maintain a constant oxygen concentration. Values in the table express mean value ± S.E.M. FCCP: carbonyl cyanide *p*-trifluoromethoxyphenylhydrazone. *t*-BOOH: *t*-butyl hydroperoxide.

active metabolizing system. The latter two requirements could be interpreted as an expression of the necessity for oxygen intermediates such as O_2^- and/or H_2O_2 . Nakano et al. (46) and Sugioka and Nakano (47) have analyzed spectroscopically the chemiluminescent signal from microsomes and liposomes undergoing lipoperoxidation. Singlet oxygen is apparently responsible for the observed chemiluminescence with light emission at 635, 580 and 520 nm (46,47); the latter two bands appear related to the quenching of singlet oxygen in a hydrophobic environment, since they do not contribute significantly to light emission in aqueous media (49,50).

Oxidized glutathione release appears at present to be more sensitive than chemiluminescence as an indicator of oxidative stress, if one compares the signals obtained from the perfused liver upon addition of *t*-butyl hydroperoxide. However, studies of chemiluminescence may lead to more suitable investigations for intact systems and, along with ethane formation (33,51,52), could allow an adequate estimation of the rates of lipid peroxidation occurring in vivo.

References

1. De Duve, C.: Physiol. Rev. 46, 323-357 (1966)
2. De Duve, C.: J. Histochem. Cytochem. 21, 941-948 (1973)
3. Sies, H.: Angew. Chem. (English ed.) 13, 706-718 (1974)
4. Poole, B.: J. Theoret. Biol. 51, 149-167 (1975)
5. Oshino, N., Jamieson, D., Chance, B.: Biochem. J. 146, 53-65 (1975)
6. Oshino, N., Chance, B.: Biochem. J. 162, 509-525 (1977)
7. Loschen, G., Flohé, L., Chance, B.: FEBS Lett. 18, 261-265 (1971)
8. Boveris, A., Chance, B.: Biochem. J. 134, 707-716 (1973)
9. Boveris, A., Oshino, N., Chance, B.: Biochem. J. 128, 617-630 (1972)
10. Gillette, J.R., Brodie, B.B., La Du, B.N.: J. Pharmacol. Exp. Ther. 119, 532-543 (1957)
11. Thurman, R.G., Ley, H.G., Scholz, R.: Eur. J. Biochem. 25, 420-430 (1972)
12. Fridovich, I.: Adv. Enzymol. 41, 35-97 (1974)
13. Hemmerich, P., Nagelschneider, G., Veeger, C.: FEBS Lett. 8, 68-93 (1970)
14. Massey, V., Strickland, S., Mayhew, S.G., Howell, L.G., Engle, P.C., Mattews, R.G., Schuman, M., Sullivan, P.A.: Biochem. Biophys. Res. Commun. 36, 891-897 (1969)
15. Misra, H.D., Fridovich, I.: J. Biol. Chem. 247, 188-192 (1972)
16. Boveris, A., Cadenas, E., Stoppani, A.O.M.: Biochem. J. 156, 435-444 (1976)
17. Cadenas, E., Boveris, A., Ragan, C.I., Stoppani, A.O.M.: Arch. Biochem. Biophys. 180, 248-257 (1977)
18. Tyler, D.D.: Biochem. J. 147, 493-507 (1975)
19. Oshino, N., Chance, B., Sies, H., Bücher, Th.: Arch. Biochem. Biophys. 154, 117-131 (1973)
20. Sies, H., Akerboom, T.P.M., Tager, J.M.: Eur. J. Biochem. 72, 301-307 (1977)
21. Mc Cord, J., Day, E.D., Jr.: FEBS Lett. 86, 139-142 (1978)
22. Kellogg, E.W., III, Fridovich, I.: J. Biol. Chem. 252, 6721-6728 (1977)
23. Pederson, T.C., Aust, S.D.: Biochim. Biophys. Acta 385, 232-241 (1975)
24. Fong, K-L., Mc Cay, P.B., Poyer, J.L., Keele, B.B., Misra, H.: J. Biol. Chem. 284, 7792-7797 (1973)
25. Cohen, G., Heikkila, R.E.: J. Biol. Chem. 249, 2447-2452 (1974)
26. Zimmermann, R., Flohé, L., Weser, U., Hartmann, H.J.: FEBS Lett. 29, 117-120 (1973)
27. Sies, H., Gerstenecker, C., Menzel, H., Flohé, L.: FEBS Lett. 27, 171-175 (1972)
28. Sies, H., Gerstenecker, C., Summer, K.H., Menzel, H., Flohé, L.: In: Glutathione. Flohé, L., Benöhr, H., Sies, H., Waller, H.D., Wendel, A. (eds.) pp. 261-275. Stuttgart: Thieme 1973
29. Sies, H., Summer, K.H.: Eur. J. Biochem. 57, 502-512 (1975)

30. Bartoli, G.M., Sies, H.: FEBS Lett. 86, 89-91 (1978)
31. Chance, B.: Circ. Res. Suppl. 38-39, 31-38 (1976)
32. Sies, H.: In: Hypoxia and ischemia. Chance, B., Coburn, R., Reivich, M., Lahiri, R. (eds.) pp. 51-66. New York: Plenum Press 1977
33. Chance, B., Saronio, C., Leigh, J.S., Jr.: Proc. Natl. Acad. Sci. U.S.A. 72, 1635-1640 (1975)
34. Riely, C.A., Cohen, G., Lieberman, M.: Science 183, 208-210 (1974)
35. Anderson, S.M., Krinsky, N.I., Stone, M.J., Clagett, D.C.: Photochem. Photobiol. 20, 65-69 (1974)
36. Gerschman, R.: In: Biological effects of oxygen. Dickens, F., Neil, E. (eds.) pp. 475-494. London: Pergamon 1964
37. Chance, B., Boveris, A.: In: Extrapulmonary manifestations of respiratory disease. Robin, E.D. (ed.) pp.185-237. New York: Marcel Dekker 1978
38. Fridovich, I.: J. Biol. Chem. 245, 4053-4057 (1970)
39. Nishiki, K., Oshino, N., Jamieson, D., Chance, B.: Biochem. J. 160, 343-355 (1976)
40. Oshino, N., Jamieson, D., Sugano, T., Chance, B.: Biochem. J. 146, 67-77 (1975)
41. Tarusov, B.N., Polivoda, A.I., Zhuravlev, A.I.: Radiobiologiya 1, 150-151 (1961)
42. Tarusov, B.N., Polivoda, A.I., Zhuravlev, A.I., Sekamova, E.N.: Tsitologiya 4, 696-699 (1962)
43. Zhuravlev, A.I., Polivoda, A.I., Tarusov, B.N.: Radiobiologiya 1, 321-324 (1961)
44. Howes, R.M., Steele, R.H.: Res. Commun. Chem. Pathol. Pharmacol. 2, 619-626 (1971)
45. Howes, R.M., Steele, R.H.: Res. Commun. Chem. Pathol. Pharmacol. 3, 349-357 (1972)
46. Nakano, M., Noguchi, T., Sugioka, K., Fujuyama, H., Sato, M., Shimizu, Y., Tsuji, Y., Inaha, H.: J. Biol. Chem. 250, 2404-2406 (1975)
47. Sugioka, K., Nakano, M.: Biochem. Biophys. Acta 423, 203-216 (1976)
48. Hochstein, P., Ernster, L.: Biochem. Biophys. Res. Commun. 12, 388-394 (1963)
49. Khan, A.U., Kasha, M.: J. Chem. Phys. 39, 2105-2106 (1963)
50. Selinger, H.H.: J. Chem. Phys. 40, 3133-3134 (1963)
51. Kappus, H., Köster, U., Köster-Albrecht, D., Kieczka, H., Remmer, H.: This volume
52. Wendel, A., Feuerstein, S., Konz, K.H.: This volume

The Selenium Moiety of Glutathione Peroxidase

A. WENDEL, B. KERNER, and K. GRAUPE

Summary

Glutathione peroxidase (E.C.1.11.1.9) the tetrametric enzyme which contains 4 atoms of selenium, was purified to homogeneity from bovine red cells and investigated by means of chemical modification and ultraviolet difference spectroscopy. The borohydride-reduced enzyme was alkylated with labeled haloacetates and the stoichiometry of binding was determined.

Chloroacetate turned out to be a suitable active-site-directed reagent. Consequently, the reduced enzyme was carboxymethylated using 1-[^{14}C]-chloroacetate, hydrolyzed, and subjected to amino acid analysis. Distribution of radioactivity and selenium resulted in a pattern similar to that obtained with analogously prepared carboxymethyl-seleno-cysteine.

The difference spectrum of substrate-reduced against oxidized GSH-peroxidase shows a distinct peak at 237 nm which is attributed to four cysteine residues per mol of enzyme. In the presence of urea, which reversibly inactivated the enzyme, KBH_4-reduced and GSH-reduced enzyme yield a difference spectrum with a sharp peak at 248 nm, corresponding to a selenol content of four groups per tetramer.

Taken together, the present findings demonstrate that the selenium moiety of GSH-peroxidase consists of a compound which can be reduced to seleno-cysteine by KBH_4 but not by GSH. In addition, a cysteine residue seems to be involved in catalysis.

A. Introduction

Glutathione peroxidase (GSH-Px) from bovine erythrocytes contains 4g-atoms of tightly bound selenium (5,11). The enzyme plays a dominant role in the defense of the cell against oxidative challenge by its ability to reduce a large variety of harmful hydroperoxides to the corresponding hydroxy compounds. Although the molecular properties (4), including X-ray analysis of the protein (7,8), have been thoroughly studied, only limited progress has been achieved in elucidating the chemical nature of the enzyme's selenium moiety: X-ray photoelectron spectroscopy (3), removal of the selenium from the oxidized enzyme (10) and notoriously preliminary observations (2,12,13) did not allow the deduction of a definite chemical structure of the selenium.

The present paper reports chemical and spectroscopic evidence that the selenium-containing compound of GSH-Px can be reduced within the protein molecule to selenocysteine.

B. Materials and Methods

Glutathione peroxidase was isolated in a modified procedure (15) using chromatography on DEAE-Sephadex, G-150, hydroxylapatite and the macroreticular resin XAD 8 (Serva, Heidelberg). Selenium was determined fluorometrically (15). Amino acid analysis was performed under N_2-atmosphere with a Unichrom analyzer on a Beckman M 82^+ resin using standard systems. Analytical runs were followed by a preparative run without ninhydrin staining. Modification experiments were carried out under N_2 throughout.

1-[^{14}C]-chloroacetate (Amersham-Buchler, specific radioactivity: 50 mCi/mmol) was used in a 10^4 times excess to selenium, 2-[^{14}C]-iodoacetate (57 mCi/mmol) 10^3 times excess, 1-[^{14}C]-iodoacetamide (57 mCi/mmol) 10^3 times excess; the reaction mixture was incubated in 0.1 M MOPS pH = 7.3. X-ray photoelectron spectra were recorded as described in (14). The protein was hydrolyzed for 15 h at 110°C in HCl under N_2.

Se-carboxymethylselenocysteine was synthezised according to (9). Cystine, methionine, selenocystamine, glycine, and histidine were subjected to similar treatment in order to obtain reference compounds for the amino acid analysis. Difference spectra were recorded in a Shimadzu UV-210 double beam spectrometer in tandem quartz cuvettes at 25°C under strict exclusion of O_2. GSH-Px was reduced by 1 mM GSH and then eluted salt-free from a G-25 column immediately into the cuvette under N_2. The contents of the two cuvette compartments are indicated in the figures.

C. Results

I. Difference Spectroscopy

Figure 1 shows the difference spectrum between 210 and 320 nm of autoxidized GSH-Px against GSH-reduced enzyme. This spectrum changed its fine structure between 275 and 300 nm upon addition of t-butylhydroperoxide to the autoxidized enzyme (not shown). The shape and the absorption of the peak at 237 nm remained the same.

Cysteine absorbs at this pH at 237 nm with an $\varepsilon = 4467$ (1). Assuming that the difference signal at this wavelength is due to cysteine, a concentration of 4.1 mol of thiol per tetramer is calculated.

Figure 2 shows that in the presence of urea an additional peak arises: the difference spectrum of KBH_4-reduced enzyme and GSH-reduced enzyme shows a sharp extremum at 248 nm, a wavelength at which selenocysteine is known to absorb with an $\varepsilon = 5900$ (3). Taking this value, a selenol concentration of four groups per tetramer is calculated. Obviously, in contrast to GSH, the strong reductant KBH_4 is able to reduce the selenium moiety to a selenol provided that the enzyme has been inactivated by urea.

II. Chemical Modifications

The stoichiometry of the binding of labeled haloacetates to the reduced enzyme was determined. Table 1 shows that during all modifications the selenium content remained unchanged. 2 mol of chloroacetate or iodoacetate were bound in the presence or absence of urea. A control experiment

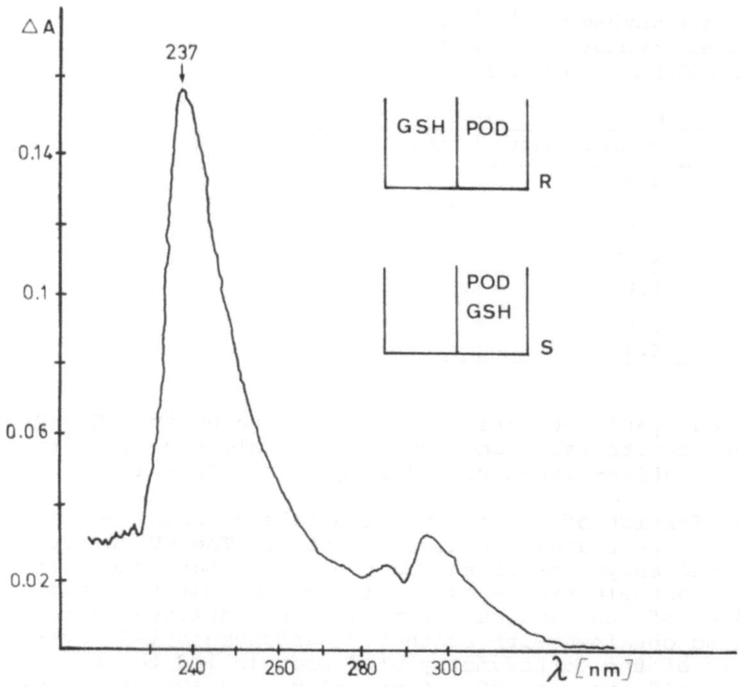

Fig. 1. Difference spectrum of native GSH-Px (POD) and enzyme in the presence of GSH. Conditions: 1.83 mg/ml GSH-Px (500 U/mg) = 6.1×10^{-5} M; 2 mM GSH; 100 mM potassium phosphate buffer pH = 7.1; d = 0.5 cm, slit = 1 nm, t = 25°. S, sample; R, reference

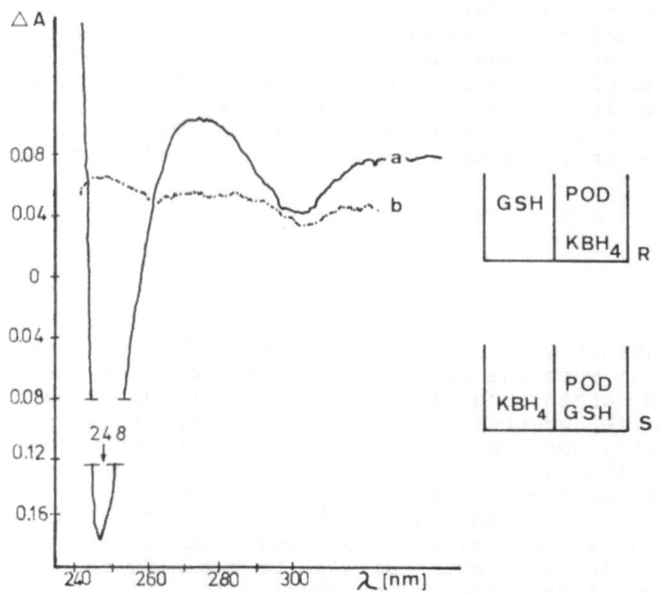

Fig. 2. a Difference spectrum of GSH-reduced and KBH_4-reduced GSH-Px (POD), b after addition of 10 μM t-butylhydroperoxide to the enzyme in the reference cuvette. Conditions: 1.1 mg/ml GSH-Px; 2 mM GSH; 2 mM KBH_4 initially; 6 M urea; 50 mM potassium phosphate buffer pH = 7.1; d = 0.5 cm; slit = 5 nm; t = 25°. S, sample; R, reference cuvette

Table 1. Stoichiometry of the binding of [^{14}C]-labeled haloacetates to native, urea-pretreated or cyanide-treated GSH-Px following reduction with KBH_4. Conditions: see Methods

Reagent	Treatment	Bound reagent	Selenium
Chloroacetate	-	2.2	4
	Urea	2.2	4
	Cyanide	0.1	0.02
Iodoacetate	-	2.1	3.5
	Urea	2.1	3.5
Iodoacetamide	-	0.1	4
	Urea	4.0	4

showed that to oxidized cyanide-treated and hence selenium-free GSH-Px (10) no labeled chloroacetate was bound. Therefore, chloroacetate was regarded as a suitable active-site-directed reagent for GSH-Px.

Before and after modification of the enzyme with labeled chloroacetate the specific selenium content again remained constant. The UV spectrum of the carboxymethylated enzyme still showed the 237 nm band under anaerobiosis. The X-ray photoelectron spectrum indicated that the original Se-3d-signal of 54.4 eV of the reduced enzyme (14) had shifted to 56.8 eV. This value was also obtained with synthetic carboxymethyl-Se-cysteine. Acid hydrolysis of the carboxymethylated protein led to losses at 70% in total amino acid content, 70% of selenium, and 40% of radioactivity. The hydrolysate was subjected to analytical and preparative amino acid analysis.

Figure 3 shows the distribution of radioactivity and selenium together with the positions of the amino acids. Obviously, the major coincidence between selenium and radioactivity lies at a retention time of 57-60 min, a position where synthetic Se-carboxymethyl-selenocysteine was eluted. In both cases (the GSH-Px hydrolysate as well as the synthetic compound) additional peaks in the region of 20-34 min were found, although not in the same quantitative relation. At 34 min, Se-carboxymethyl-selenocysteine-selenon is eluted, one of the expected autoxidation products.

D. Discussion

The fact that GSH-Px, together with its donor substrate GSH, shows a difference signal at 237 nm when the reference cuvette contains untreated enzyme allows two conclusions: (1) a functional group has been reduced by GSH in the sample cuvette; (2) the enzyme is not fully reduced in the "native" state, not even after previous reduction under careful exclusion of O_2. Although amino acid analysis of the performic acid oxidized enzyme revealed eight cysteinic acids per mol (6,15), the absorption at 237 nm is equivalent to only four cysteine residues. Since this absorption needs several minutes to appear, it is concluded that reducible cysteine reflects an event similar to the slow reactivation by GSH of the enzyme first observed by Günzler (6).

Surprisingly a new peak at 248 nm arises if reduction by KBH_4 is performed in urea. Controls with selenocysteine and cysteine in urea showed that it is not the solvent that might have changed either ab-

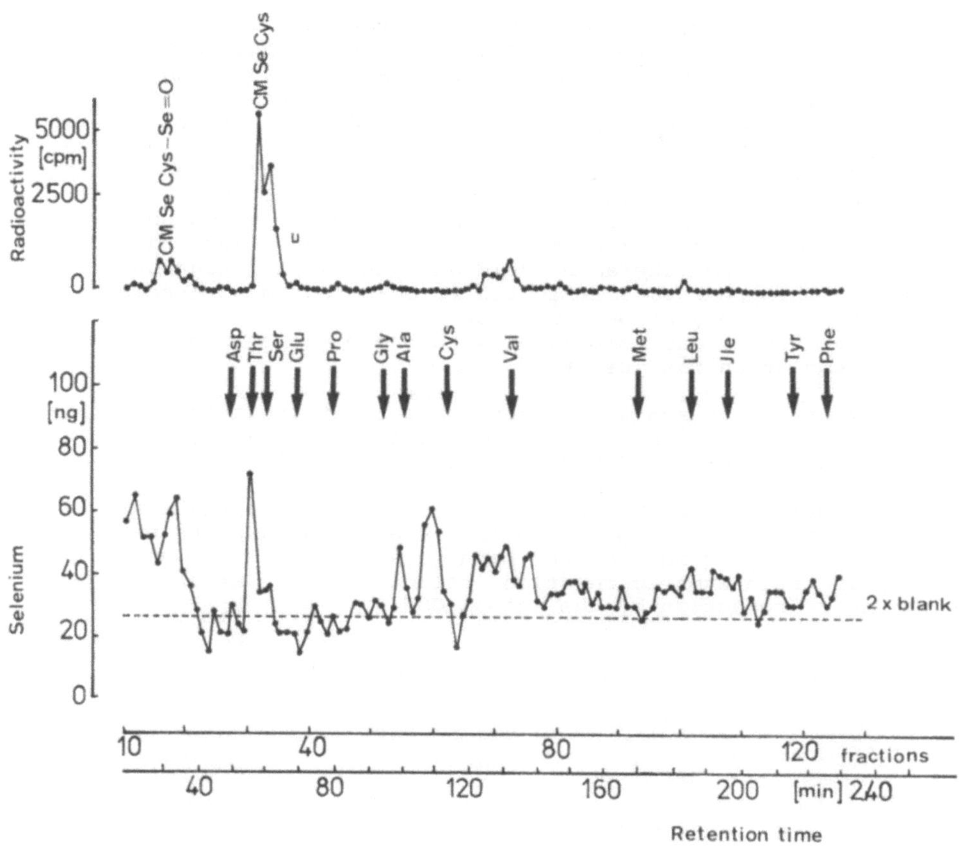

Fig. 3. Amino acid analysis, distribution of radioactivity and selenium of [^{14}C]-carboxymethylated GSH-Px after acidic total hydrolysis. Arrows indicate the retention times at which the amino acids were eluted in this system

sorption. Thus this selenol peak really emerges additionally. Obviously, this group participates in catalysis, since it disappears within seconds upon addition of t-butyl-hydroperoxide, the acceptor substrate, and can be restored by excess GSH.

The finding that 2 mol chloroacetate or iodoacetate were bound to the tetrameric GSH-Px concomitant with complete inactivation is difficult to explain. Two interpretations are offered. (1) There are indeed other oligomeric enzymes where only one half of the molecule reacts. (2) If an instable intermediate like a carboxymethylselenonium is involved, its decomposition leads to labeled hydroxyacetate which is removed with the low-molecular weight fraction, while the other half remains attached to the protein. Nevertheless, carboxymethylation with chloroacetate showed pseudo-zero-order kinetics with a rate constant of 1.2 min^{-1}. The selenium recovery and balance proved to be satisfactory. The control experiment with cyanide-treated oxidized enzyme showed that the presence of selenium is required for the binding of chloroacetate. The binding energy of the selenium 3 d-electrons as measured by the X-ray photoelectron spectrum indicates a redox change of the selenium atom and with 56.8 eV a position between the values for the reduced (54.1 eV) and the oxidized (58.0 eV) enzyme (14). Together these data seem to justify the statement that selenium within the active-site had indeed been chemically modified.

The losses of selenium, radioactivity, and protein could not be minimized in several trials. Since the ratio of total amino acid content to selenium was constant, it seemed appropriate to regard the coincidence of selenium and radioactivity as positions where products of the active-site-directed modification are eluted. Unfortunately, the major peak of selenium + radioactivity is covered by the presence of 64 aspartyl residues. Therefore a quantification via ninhydrin was not possible. In spite of considerable experimental efforts, the formation of autoxidation products of the carboxymethylated selenium moiety could not be prevented. In many experiments, however, the major modification product was eluted from the amino acid analyzer with retention times of 57-60 min. This peak was attributed to Se-carboxymethyl-selenocysteine. Rechromatography of one fraction of the analytical run in a much more sensitive system and comparison with the model compound showed that this association was justified.

Both experimental approaches used here demonstrate that free selenocysteine does not occur in GSH-peroxidase. The X-ray analysis showed that the selenium atoms of GSH-Px have a distance of 21 Å within a dimer and 39.5 Å between the two dimers (8). Thus a diselenide is excluded. The spectral evidence favors an intramolecular sulfoselenide R-Cy-S-Se-Cy-R' between a protein-bound selenocysteine and a protein-bound cysteine. Thus, replacement of the bound cysteine by GSH would be the first step in the enzymatic reaction, followed by oxidation of the selenium and re-reduction by the second GSH.

Acknowledgments. This work was financially supported by the Deutsche Forschungsgemeinschaft (Grant We 686/3).

References

1. Benesch, R.E., Benesch, R.: The acid strength of the -SH group in cysteine and related compounds. J. Am. Chem. Soc. 77, 5877-5881 (1955)
2. Chiu, D., Fletcher, B., Stults, F., Zakowski, J., Tappel, A.L.: Properties of selenium-glutathione peroxidase. Federation Proc. 34, 925 (1975)
3. Dickson, A.C., Tappel, A.L.: Reduction of selenocystine by cysteine or glutathione. Arch. Biochem. Biophys. 130, 547-550 (1969)
4. Flohé, L., Eisele, B., Wendel, A.: Glutathion-Peroxidase I: Reindarstellung und Molekulargewichtsbestimmungen. Hoppe-Seyler's Z. Physiol. Chem. 352, 151-158 (1971)
5. Flohé, L., Günzler, W.A., Schock, H.H.: Glutathione peroxidase: A selenoenzyme. FEBS Lett. 32, 132-134 (1973)
6. Günzler, W.A.: Glutathion-Peroxidase. Kristallisation, Selengehalt, Aminosäurezusammensetzung und Modellvorstellungen zum Reaktionsmechanismus. Dissertation Univ. Tübingen, 1974
7. Ladenstein, R., Wendel, A.: Crystallographic data of the selenoenzyme glutathione peroxidase. J. Mol. Biol. 104, 877-882 (1976)
8. Ladenstein, R., Wendel, A.: unpublished (1978)
9. de Marco, C., Rinaldi, A., Dessi, M.R., Dernini, S.: Oxidation of Se-carboxymethyl-selenocysteine by L-amino acid oxidase and by D-aspartate oxidase. Mol. Cell. Biochem. 12, 92-98 (1976)
10. Prohaska, J.R., Oh, S.H., Hoekstra, W.G., Ganther, H.E.: Glutathione peroxidase: Inhibition by cyanide and release of selenium. Biochem. Biophys. Res. Commun. 74, 64-71 (1977)
11. Rotruck, J.T., Pope, A.A., Ganther, H.E., Swanson, S.B., Hafemann, D., Hoekstra, W.G.: Selenium: Biochemical role as a component of glutathione peroxidase. Science 179, 588-590 (1973)

12. Stults, F.H., Chiu, D., Fletcher, B., Forstrom, J., Hsu, C., Pierce, S., Tappel, A.L.: Properties of purified glutathione peroxidase and its selenium prosthetic group. Federation Proc. 35, 860 (1976)
13. Tappel, A.L., Forstrom, J.W., Zakowski, J.J., Lyons, D.E., Hawkes, W.C.: The catalytic site of rat liver glutathione peroxidase as selenocysteine and selenocysteine in rat liver. Federation Proc. 37, 706 (1978)
14. Wendel, A., Pilz, W., Ladenstein, R., Sawatzki, G., Weser, U.: Substrate-induced redox-change of selenium in glutathione peroxidase studied by X-ray photoelectron spectroscopy. Biochim. Biophys. Acta 377, 211-215 (1975)
15. Wendel, A., Kerner, B.: unpublished (1978)

Non-Selenium-Dependent Glutathione Peroxidase*

R. F. BURK and R. A. LAWRENCE

Summary

Recently a non-selenium-dependent glutathione peroxidase activity (non-Se GSH-Px) has been recognized in rat liver in addition to the selenium-dependent glutathione peroxidase (Se GSH-Px). Work by several groups indicates that non-Se GSH-Px is due to GSH S-transferase B and possibly GSH S-transferases A and C. Kinetic studies of GSH S-transferase B reveal the Km of its non-Se GSH-Px activity with cumene hydroperoxide to be 0.55 mM and that with t-butyl hydroperoxide to be 2.3 mM. Non-Se GSH-Px will not utilize 0.25 mM H_2O_2 as a substrate. In contrast Se GSH-Px utilizes all these substrates and Km's are in the range of 10-50 μM. Non-Se GSH-Px is found in rat liver, kidney, testis, adrenal, brain, and fat. It is present in liver from hamster, rat, sheep, pig, chicken, human being, and guinea pig. Because of its unfavorable Km's when compared with Se GSH-Px, the role of non-Se GSH-Px is uncertain. However, it increases in the rat liver in selenium deficiency; and, using a hemoglobin-free liver perfusion system, it has been shown to remove organic hydroperoxides in a selenium-deficient liver. Thus, it may function in peroxide metabolism under some conditions.

A. Introduction

Investigations of selenium-dependent glutathione peroxidase (Se GSH-Px), measured using H_2O_2 as substrate, indicate that it is virtually undetectable in the livers of selenium-deficient rats. We became suspicious that a second GSH-Px existed when we noted GSH-Px activities as high as 30% of control in selenium-deficient rat livers when cumene hydroperoxide (CuOOH) was used as substrate. Our subsequent studies of this activity (Lawrence and Burk, 1976), led to recognition of the existence of non-selenium-dependent glutathione peroxidase (non-Se GSH-Px).

The gel filtration experiment with hepatic 105,000 g supernatant shown in Figure 1 demonstrates the two GSH-Px activities. Peak I, which is the Se GSH-Px, disappears in selenium deficiency. It uses both H_2O_2 and CuOOH as substrate. Peak II, which elutes at molecular weight 39,000, does not decrease in selenium deficiency and will not use H_2O_2 in the usual concentrations as substrate. Thus, peak II is the non-Se GSH-Px and is responsible for the residual GSH-Px found in selenium-deficient liver supernatant when CuOOH is used as substrate. Other investigators apparently observed this activity in brain (Prohaska and Ganther, 1976), but did not carry out the investigations necessary to characterize it as a non-Se GSH-Px.

* Supported by NIH grant ES 01992.

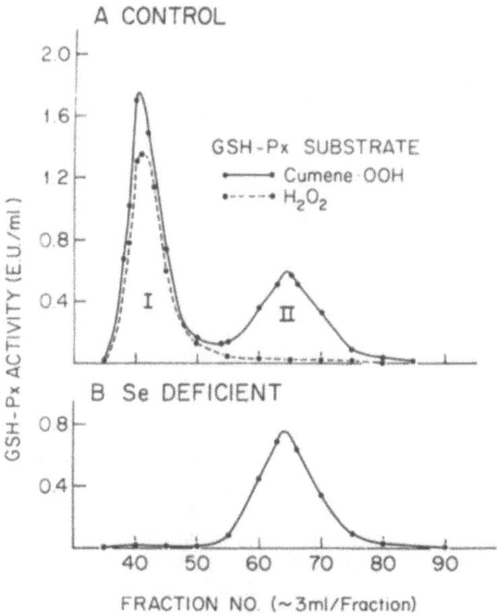

Fig. 1A and B. Sephadex G-150 gel filtration of rat liver 105,000 g supernatant (Lawrence and Burk, 1976). Substrate concentrations were: 1.5 mM cumene hydroperoxide and 0.25 mM H_2O_2

B. Purification and Properties

A major part if not all of the non-Se GSH-Px of rat liver 105,000 g supernatant is due to some of the GSH S-transferases. The physical properties of these proteins and the characteristics of their glutathione S-transferase reactions have been reviewed recently (Jakoby, 1978), and are also discussed elsewhere in these proceedings. Using carboxymethyl cellulose ion exchange chromatography, Prohaska and Ganther (1977) showed coincidence of GSH S-transferase activity with GSH-Px in three peaks which they felt represented GSH S-transferases AA, B, and C based on comparison with similar chromatograms published by others. In addition they observed a GSH-Px peak unassociated with GSH S-transferase activity as assayed with the substrate 1-chloro-2,4-dinitrobenzene. They studied some of the kinetic characteristics of their major GSH-Px peak which corresponded to GSH S-transferase B.

Lawrence et al. (1978) published carboxymethyl cellulose chromatography data similar to those of Prohaska and Ganther, which they felt indicated GSH-Px activity in peaks corresponding to GSH S-transferases A, B, and C. However, since the transferase peaks eluted from carboxymethyl cellulose are mixtures of transferases (Hale and Neims, 1976), and others have employed a hydroxylapatite step to obtain homogeneous preparations (Habig et al., 1974), Lawrence et al. used such a step with their GSH S-transferase B peak from the carboxymethyl cellulose column. They then carried out kinetic and inhibition studies with this preparation.

Pierce and Tappel (1978) also purified a fraction from rat liver having both GSH S-transferase and GSH-Px activities and studied its kinetic properties. They did not, however, characterize their preparation with respect to the usual classification of GSH S-transferases and it is unclear whether it represents one transferase or a mixture.

Table 1. Km values

Preparation	Substrate	Km (mM)
Prohaska and Ganther (1977)	CuOOH[a]	0.19, 0.17
	BuOOH[a]	5.0 , 5.7
	1-chloro-2,4-dinitrobenzene[a]	0.9 , 0.3
Lawrence et al. (1978)	CuOOH	0.55
	BuOOH[b]	2.32
Pierce and Tappel (1978)	CuOOH	0.57
Habig et al. (1974)	1-chloro-2,4-dinitrobenzene	0.8
	sulfobromophthalein	0.07
	1,2-dichloro-4-nitrobenzene	2.0

[a] Apparent Km at 1 mM GSH. [b] Apparent Km at 1.2 mM GSH

Table 1 presents Km values resulting from the studies of these groups of the respective GSH S-transferase preparations functioning as a GSH-Px and compares them with some of the published Km's of GSH S-transferase B. The Km's of both modes of function are in the same range. The Km's for hydroperoxides with this activity, however, are much higher than they are with Se GSH-Px (Günzler et al., 1972). Non-Se GSH-Px showed zero-order kinetics with respect to GSH and an initial velocity pattern indicative of an ordered sequential reaction (Prohaska and Ganther, 1977; Lawrence et al., 1978; Pierce and Tappel, 1978). This contrasts with the first-order kinetics with respect to GSH for the Se GSH-Px and a proposed ping pong mechanism (Günzler et al., 1972).

Lawrence et al. (1978) showed that sulfobromophthalein is a competitive inhibitor of the non-Se GSH-Px. Since this compound is a known substrate of GSH S-transferase B, this provides further evidence of the identity of these enzymes and suggests that the binding site for substrates may be the same for both activities. Thus, it appears that much of the non-Se GSH-Px of rat liver is due to GSH S-transferase B. Whether other GSH S-transferases have GSH-Px activity has not yet been established.

C. Occurrence

Non-Se GSH-Px is present in a number of tissues of the rat. A recent survey (Lawrence and Burk, 1978), which identified Se GSH-Px and non-Se GSH-Px by their elution volumes on gel filtration chromatography, yielded the findings in Figure 2. Heart, red blood cells, skin, skeletal muscle, spleen, lung, thymus, and intestinal mucosa were also examined and no non-Se GSH-Px was found in them. They all contained Se GSH-Px. Testis was the only tissue of those studied which contained more non-Se GSH-Px than Se GSH-Px. Others (Prohaska and Ganther, 1977) have reported lower ratios of non-Se GSH-Px to Se GSH-Px in rat livers, kidney, brain, and testis than those shown in Figure 2. Their results are explained by their use of lower CuOOH concentrations in the assay, which are not saturating for the non-Se GSH-Px, but which do saturate the Se GSH-Px.

GSH S-transferase B has been identified immunochemically in rat liver, kidney, and intestinal mucosa (Jakoby, 1978). Failure to find non-Se GSH-Px in intestinal mucosa, then, was likely due to technical problems such as its inactivation by digestive enzymes during tissue preparation.

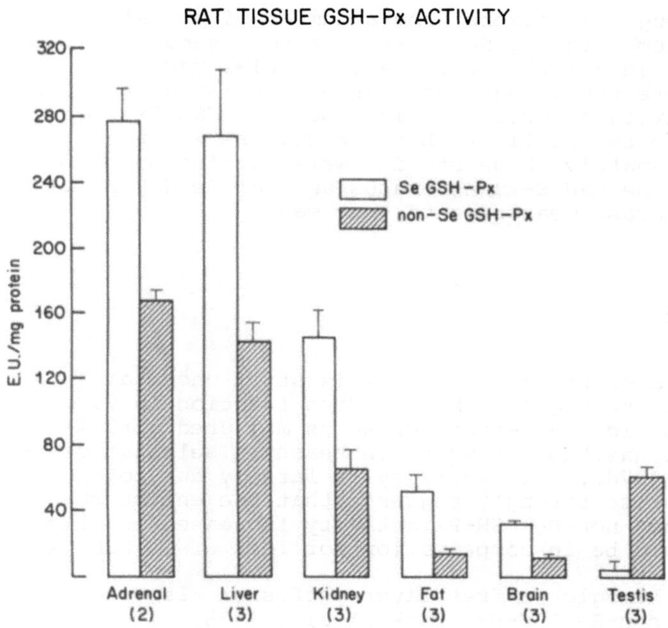

Fig. 2. Distribution of Se GSH-Px and non-Se GSH-Px in 105,000 g supernatants of rat tissues. Data are means + S.E. of the number of observations shown in parentheses and are from Lawrence and Burk (1978). Experimental details may be found there. Substrate was 1.5 mM cumene hydroperoxide

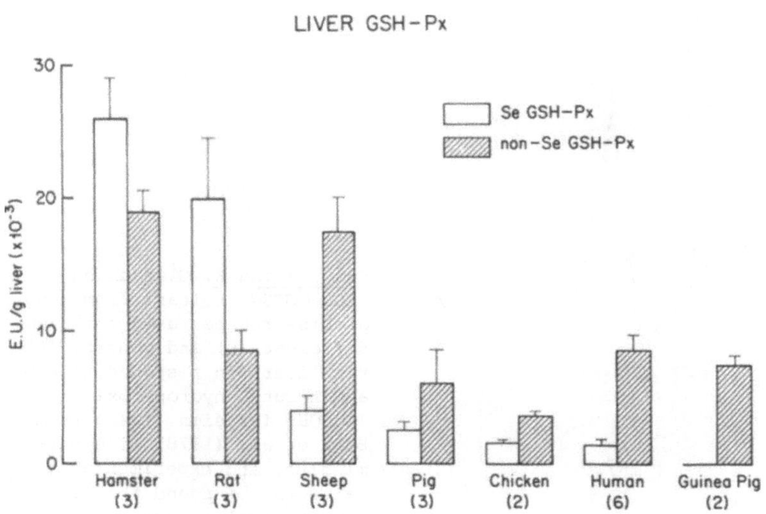

Fig. 3. Distribution of Se GSH-Px and non-Se GSH-Px in 105,000 g supernatants of livers from a variety of animals. Data are means + S.E. of the number of observations shown in parentheses and are from Lawrence and Burk (1978). Experimental details may be found there. Substrate was 1.5 mM cumene hydroperoxide

Non-Se GSH-Px has been sought in livers from a number of animal species and detected in all of them (Fig.3). Se GSH-Px was not found in guinea pig liver but was present in all the rest. Large species variations in both GSH-Px activities were noted and there was no consistent relationship between the two activities across species. More Se GSH-Px than non-Se GSH-Px was found in rodent liver, but the reverse was seen in the other species. Unfortunately these studies were carried out before the relationship between the GSH S-transferases and non-Se GSH-Px was known and no GSH S-transferase measurements were made.

D. Physiological Function

The physiological role played by non-Se GSH-Px is still uncertain. However, several observations suggest that it does function in vivo. First the activity of hepatic GSH S-transferase as measured with 1-chloro-2,4-dinitrobenzene has been shown to increase in selenium deficiency (Lawrence et al., 1978). This activity is largely due to GSH S-transferase B, and the rise strongly suggests that the enzyme responsible for at least some non-Se GSH-Px activity increases in selenium deficiency. This could be in compensation for loss of Se GSH-Px.

Studies carried out with hemoglobin-free liver perfusion also provide evidence for function of non-Se GSH-Px (Burk et al., 1978). In those studies the system of Sies et al. (1972), in which effluent glutathione is used as an index of GSH-Px function, was employed with selenium-deficient and control liver. Selenium-deficient livers contained less than 5% of control Se GSH-Px and could not release glutathione in response to H_2O_2 infusion in contrast to control livers (Fig.4). However, both selenium-deficient and control livers released glutathione in

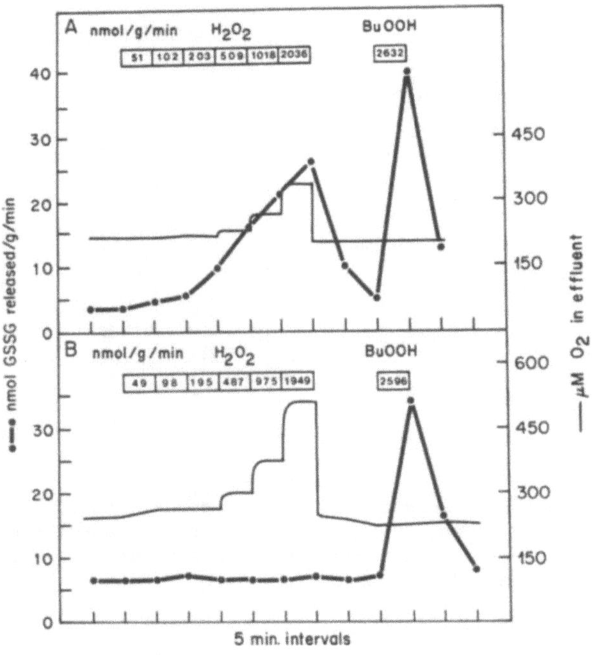

Fig. 4A and B. Glutathione (GSH + GSSG) release from hemoglobin-free perfused selenium-deficient (B) and control (A) rat livers in response to H_2O_2 and t-butyl hydroperoxide (BuOOH) infusion. Taken from Burk et al. (1978) with permission, and experimental details may be found there. Effluent samples were taken at 5-min intervals just before a change in H_2O_2 infusion rate. Infusion rates and duration are shown in the boxes at the top of each panel. Perfusate O_2 was 740 µM and effluent O_2 is shown in the figure response to t-butyl-hydroperoxide (BuOOH) infusion. Since non-Se GSH-Px can utilize BuOOH as a substrate (Table 1), but cannot utilize H_2O_2, this result is taken as evidence of function of non-Se GSH-Px in the perfused liver.

There appears to be little chance that non-Se GSH-Px functions in H_2O_2 removal. It failed to utilize 0.25 mM H_2O_2 as substrate and to cause glutathione release from perfused liver when H_2O_2 was infused. Because of the unfavorable Km's with CuOOH and BuOOH when compared with Se GSH-Px, the non-Se GSH-Px would appear to have a minor role in the removal of such substrates when they are present in low (<0.1 mM) concentrations and when Se GSH-Px is present. However, when Se GSH-Px is not present, such as in guinea pig liver, non-Se GSH-Px function may be important. Furthermore, natural peroxide substrates for GSH-Px may be better substrates for non-Se GSH-Px than the artificial ones tested so far. Thus, elucidation of the physiological role of non-Se GSH-Px and its relationship with Se GSH-Px remains largely a task for the future.

References

Burk, R.F., Nishiki, K., Lawrence, R.A., Chance, B.: Peroxide removal by selenium-dependent and selenium-independent glutathione peroxidases in hemoglobin-free perfused rat liver. J. Biol. Chem. 253, 43-46 (1978)

Günzler, W.A., Vergin, H., Müller, I., Flohé, L.: Glutathion-Peroxidase VI. Die Reaktion der Glutathion-Peroxidase mit verschiedenen Hydroperoxiden. Hoppe-Seyler's Z. Physiol. Chem. 353, 1001-1004 (1972)

Habig, W.H., Pabst, M.J., Jakoby, W.B.: Glutathione S-transferases. J. Biol. Chem. 249, 7130-7139 (1974)

Hale, B.F., Neims, A.H.: A sex difference in hepatic glutathione S-transferase B and the effect of hypophysectomy. Biochem. J. 160, 223-229 (1976)

Jakoby, W.B.: The glutathione S-transferases: A group of multifunctional detoxification proteins. In: Advances in enzymology. Meister, A. (ed.), Vol.46, pp.383-414. New York: John Wiley 1978

Lawrence, R.A., Burk, R.F.: Glutathione peroxidase activity in selenium-deficient rat liver. Biochem. Biophys. Res. Commun. 71, 952-958 (1976)

Lawrence, R.A., Burk, R.F.: Species, tissue and subcellular distribution of non-Se-dependent glutathione peroxidase activity. J. Nutr. 108, 211-215 (1978)

Lawrence, R.A., Parkhill, L.K., Burk, R.F.: Hepatic cytosolic non-selenium-dependent glutathione peroxidase activity: Its nature and the effect of selenium deficiency. J. Nutr. 108, 981-987 (1978)

Pierce, S., Tappel, A.L.: Glutathione peroxidase activities from rat liver. Biochem. Biophys Acta 523, 27-36 (1978)

Prohaska, J.R., Ganther, H.E.: Selenium and glutathione peroxidase in developing rat brain. J. Neurochem. 27, 1379-1387 (1976)

Prohaska, J.R., Ganther, H.E.: Glutathione peroxidase activity of glutathione-S-transferases purified from rat liver. Biochem. Biophys. Res. Commun. 76, 437-445 (1977)

Sies, H., Gerstenecker, C., Menzel, H., Flohé, L.: Oxidation in the NADP system and release of GSSG from hemoglobin-free perfused rat liver during peroxidatic oxidation of glutathione by hydroperoxides. FEBS Lett. 27, 171-175 (1972)

Properties of Glutathione Disulfide (GSSG) and Glutathione-S-Conjugate Release from Perfused Rat Liver

H. SIES, A. WAHLLÄNDER, and CH. WAYDHAS

Summary

Glutathione disulfide efflux from perfused rat liver, initially observed during metabolism of externally added hydroperoxides, was studied more recently in different conditions possibly associated with changes in endogenous hydroperoxide metabolism, including drug oxidations, phenobarbital pretreatment, and selenium deficiency. While GSSG efflux was increased during oxidative transitions like drug oxidations or hydroperoxide reduction, GSH efflux remained unchanged. Further, it was found that there are distinct pathways for GSSG and GSH efflux: GSSG is released into bile, whereas GSH is released into the caval perfusate. Thus, GSSG appears to be acted upon by the biliary excretory system as the "glutathione-S-conjugate of glutathione".

When GSSG production rates were increased, efflux was observed to occur also into the caval perfusate. Similar observations were made with the efflux of S-(2,4-dinitrophenyl)glutathione generated intracellularly by reaction of glutathione with 1-chloro-2,4-dinitrobenzene, catalyzed by the glutathione-S-transferases. The concentration of this conjugate in bile rose up to a maximum value of 24 mM, reached at a rate of conjugate formation of approximately 30 nmol/min per g of liver wet weight. A further increase in conjugate formation rate resulted in augmented caval perfusate efflux of the conjugate. Interestingly, there is an extra release of glutathione disulfide into bile during excretion of glutathione-S-conjugates.

A. Introduction

Addition of hydroperoxides to the perfused liver leads to an extra release of glutathione disulfide (GSSG) into the extracellular space (1-4) due to an acceleration of the GSH peroxidase reaction:

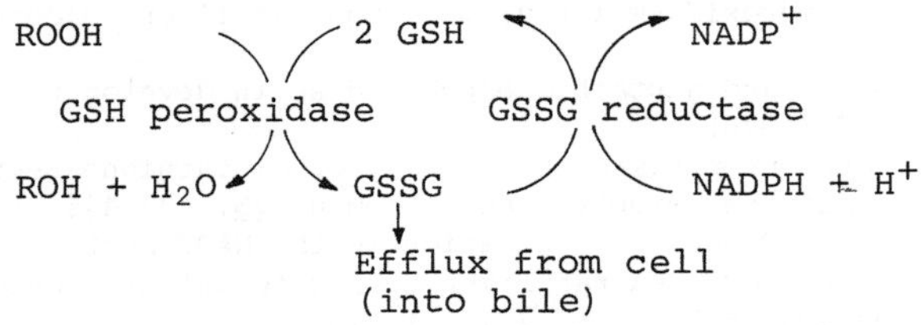

Thus, efflux of GSSG occurs even in the presence of GSSG reductase which catalyzes the reduction of disulfide at the expense of NADPH. It has been shown that NADPH in the liver cell is oxidized to NADP$^+$ upon addition of hydroperoxides, and further metabolic consequences have been described (1-5).

Since the GSSG efflux from the cell may be utilized as a sensitive indicator of endogenous hydroperoxide production (4-9), further knowledge on properties of GSSG efflux is of interest. In this paper, we will briefly describe our recent finding (10) that GSSG efflux occurs into the bile, whereas GSH, also released from the liver cell (11), is released into the caval perfusate in perfused liver. Thus, GSSG appears to be acted upon by the biliary excretory system as the "glutathione-S-conjugate of glutathione". In a second section, some properties of glutathione-S-conjugate release in the isolated perfused rat liver will be briefly presented.

B. Materials and Methods

Male Wistar rats of 150-210 g body weight were used for the liver perfusions. Phenobarbital pretreatment was performed by i.p. injection of 80 mg phenobarbital per kg body weight daily for three days prior to the experiment. The technique of hemoglobin-free, nonrecirculating perfusion of liver at 37°C with the standard bicarbonate-buffered saline solution was used as described previously (12) except that the Ca^{2+} concentration was 1.25 mM. The bile duct was cannulated, and bile was collected at 2.5 min intervals. The influent perfusate contained L-lactate (2.1 mM) and pyruvate (0.3 mM) as sodium salts. Perfusate flow was about 4 ml/min per g of liver wet weight.

Effluent caval perfusate and bile were assayed directly for glutathione disulfide by following the decrease of NADPH absorbance after addition of glutathione reductase (4); for reduced glutathione by following the formation of S-lactoyl glutathione from methylglyoxal, catalyzed by glyoxalase I (13); and for GSH + 2 GSSG (in GSH equivalents) in the catalytic assay using glutathione reductase and 5,5'-dithiobis-(nitrobenzoic acid) (5,14). The concentration of S-(2,4-dinitrophenyl)glutathione was measured according to (15).

C. Results and Discussion

I. Glutathione Disulfide Efflux from Rat Liver in Various Metabolic Conditions

As shown in Table 1, GSSG efflux from perfused liver is increased upon the addition of a hydroperoxide as well as upon addition of an H_2O_2-generating substrate, benzylammonium chloride. However, there are no significant differences from the controls if the rats are pretreated with phenobarbital or if kept on a diet low in selenium. Thus, these metabolic states are not associated with an extra hydroperoxide production detectable by GSSG release, whereas it is well-known, for example, that microsomal membranes from phenobarbital-pretreated rats have an increased capacity of H_2O_2 production. These problems are further discussed in (8).

Further, it is seen in Table 1 that the addition of drug substrates leads to an extra efflux of GSSG. The aminopyrine-dependent GSSG release is observed also in livers from selenium-deficient phenobarbital-treated rats, whereas the H_2O_2-dependent GSSG efflux is abolished under such circumstances (7) (Table 2). Thus, it was concluded (8) that these observations do not support the postulate of an extra production of free H_2O_2 during drug oxidation in the intact cell.

Table 1. Efflux of glutathione disulfide from hemoglobin-free perfused rat liver under different metabolic conditions. Data from (8,11), given as means ± S.E.M., number of observations in parentheses

Pretreatment of rats	Additions	Efflux of GSSG nmol/min per g of liver	
Control, fed	None	1.0 ± 0.1	(10)
	t-butyl hydroperoxide (0.5 μmol/min per g)	18.7	(3)
fasted, substrate-free perfused	Benzylammonium chloride (1 mM)	22.0	(3)
Selenium-deficient, fed	None	1.2	(3)
fasted, substrate-free perfused	Benzylammonium chloride (1 mM)	2.0	(3)
Phenobarbital-treated, fed	None	1.0 ± 0.1	(6)
	Aminopyrine (0.8 mM)	5.8 ± 1.3	(4)
	Ethylmorphine (0.4 mM)	2.7 ± 0.4	(4)
	Hexobarbital (0.4 mM)	4.8 ± 1.0	(4)

Table 2. GSSG efflux from phenobarbital-pretreated rat livers in different states of selenium supply. Substrate-free perfused; data are means ± S.E.M. (n=6) [from (8)]

Additions	GSSG efflux	
	Se-supplemented	Se-deficient
	nmol/min per g of liver	
None	1.2 ± 0.1	1.1 ± 0.1
Aminopyrine (0.7 mM)	3.6 ± 0.4	3.6 ± 0.4
H_2O_2 (0.16 mM)	5.7 ± 0.2	1.1 ± 0.1

The efflux of GSH from the perfused rat liver, determined by direct assay of GSH with the glyoxalase I, was not significantly affected by the oxidative transitions shown in Tables 1 and 2 (8,11), suggesting that the efflux of GSH and GSSG from the liver proceeds by distinct pathways.

II. GSSG Efflux Occurs into Bile, and GSH Occurs into Caval Perfusate

On the assumption that GSSG (MW 612) may be visualized as the "glutathione-S-conjugate of glutathione" with its molecular weight well above the critical molecular weight of approximately 325 for biliary export (16) (GSH (MW 307) is slightly below this threshold), the bile was collected separately and glutathione assays performed in simultaneous samples of bile and caval perfusate (10). As shown in Figure 1A, glutathione appears in the bile, and an increase occurs upon the addition of benzylammonium chloride or of t-butyl hydroperoxide. The rates of release of glutathione into the caval perfusate and into the bile, calculated from Figure 1A by multiplication with the perfusate and bile flow, respectively, are presented in Figure 1B. It is seen that the steady-state rates of release are approximately 10 and 2 nmol/min per g of liver, respectively. The increased rate of release into the caval perfusate upon the addition of t-butyl hydroperoxide may reflect an overloading of the biliary pathway, since it does not occur at lower

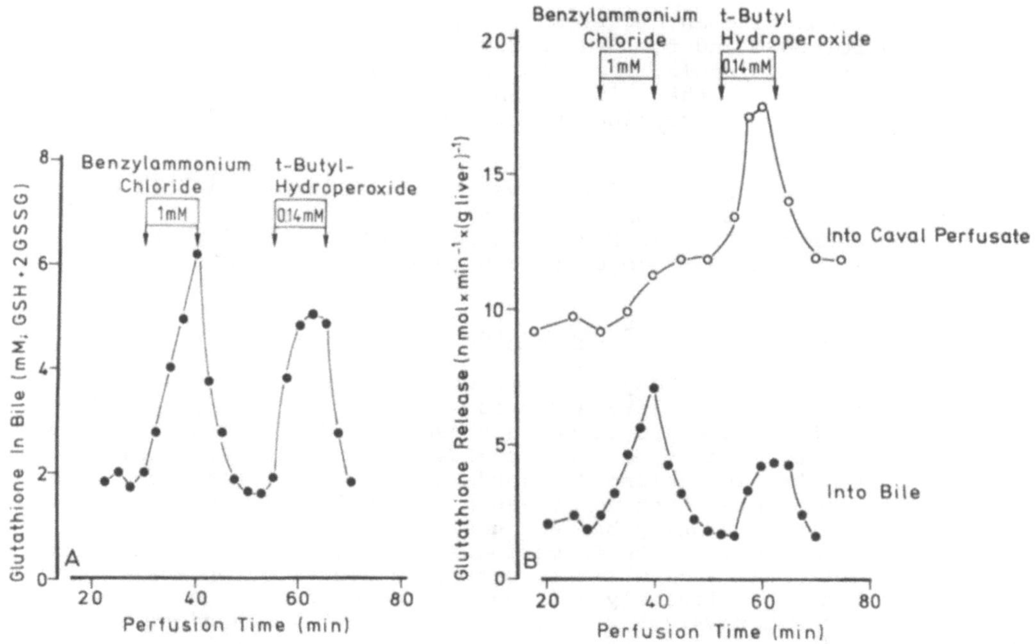

Fig. 1A and B. Efflux of glutathione from perfused liver upon the addition of benzyl-ammonium chloride and t-butyl hydroperoxide. (A) The concentration of glutathione in the bile, assayed by the catalytic test (14). (B) Rates of glutathione release into caval perfusate and into bile

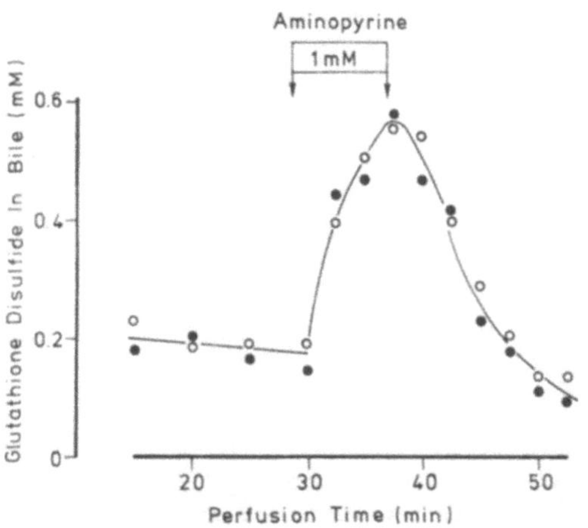

Fig. 2. Efflux of GSSG from perfused liver of phenobarbital-pretreated rat and its stimulation upon addition of a drug substrate, aminopyrine. Glutathione was assayed as GSSG (o) and as GSH + 2 GSSG (•). Since the results of the two assay methods are similar, indicating no release of GSH into the bile, both are presented on the scale as GSSG

rates of infusion of t-butyl hydroperoxide. In addition, there is a slight decrease in bile flow from 1.0 µl/min per g of liver to 0.8 µl/min per g of liver after the addition of the hydroperoxide in the experiment of Figure 1.

The increased release of GSSG upon addition of drug substrate (Table 1) is reflected by an increase in the GSSG concentration in bile. This is shown for aminopyrine in Figure 2. Here, the samples have been assayed separately for GSSG and for GSH + 2 GSSG, and the results are shown by the two different symbols. The data indicate similar values for both types of assay, i.e., there is no GSH in the bile. Similar experiments carried out with ethylmorphine and with hexobarbital as drug substrates gave similar results.

A comparison of Figure 1A and 2 shows an approximate five-fold difference in the initial GSSG concentration, assuming that all of the glutathione found in the bile in Figure 1A was GSSG. In ten different perfusions, the biliary GSSG concentration was 0.45 ± 0.23 mM.

III. Efflux of Glutathione-S-Conjugates from Perfused Liver

Since the glutathione-S-conjugates in general have the physicochemical properties that favor biliary excretion (16,17), we have compared some properties of S-conjugate release to those of GSSG release. As shown in Figure 3, the infusion of 1-chloro-2,4-dinitrobenzene, a well-known substrate for the glutathione-S-transferases (18,19), leads to formation of the conjugate which is released into bile and into the caval perfusate. The partitioning between these two pathways is dependent upon the rate of S-conjugate formation. The conjugate is released predominantly into bile at low rates, but conjugate also appears in the caval perfusate, particularly at higher rates of conjugate formation (Figure 3). In Figure 4, the S-(2,4-dinitrophenyl)glutathione concentration in bile is shown as a function of the rate of infusion of 1-chloro-2,4-dinitrobenzene; the maximal biliary concentration of the S-conjugate is 24 mM.

It is of interest to note, in addition, that the glutathione-S-conjugate release into bile is accompanied by an extra release of GSSG into the bile (Fig.5). This coincidence of S-conjugate and GSSG release may be of significance for the mechanism of the biliary secretion of the conjugates (A. Wahlländer, H. Sies, in prep.).

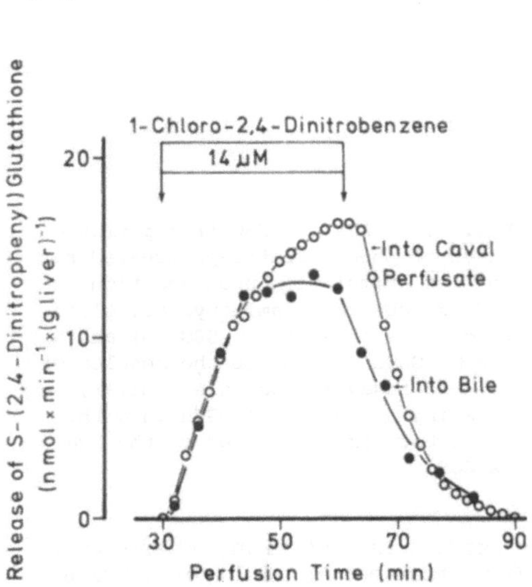

Fig. 3. Release of the glutathione-S-conjugate from the perfused liver into bile and caval perfusate. The conjugate is formed intracellularly by reaction of added 1-chloro-2,4-dinitrobenzene with glutathione

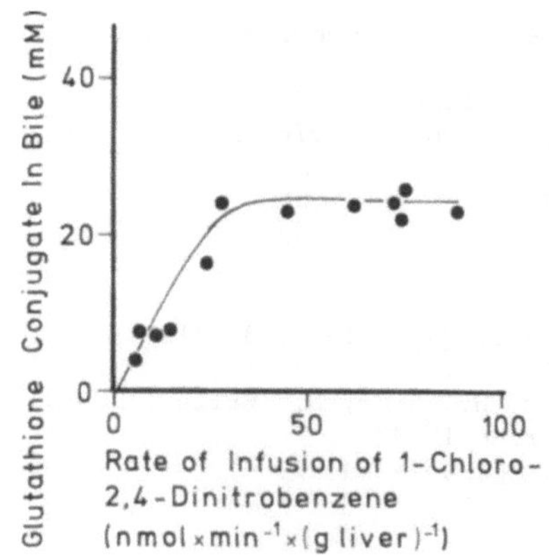

Fig. 4. Saturation curve for S-(2,4-dinitrophenyl)glutathione concentration in bile

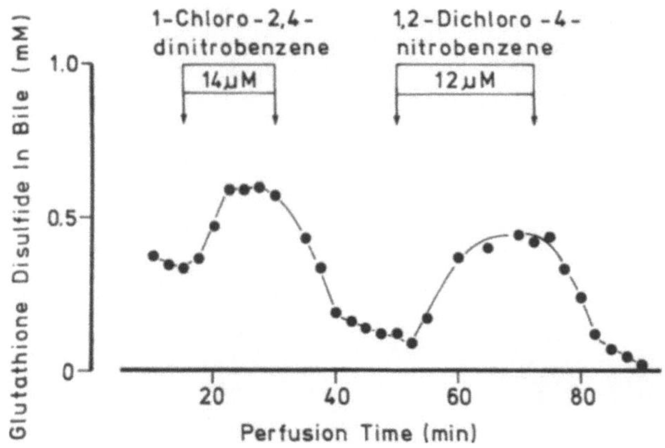

Fig. 5. Glutathione disulfide efflux into bile stimulated during infusion of substrates for glutathione-S-transferases (cf.Fig.3)

Acknowledgments. Expert technical assistance was provided by Annegret Marklstorfer and Ingrid Linke. These studies were supported by Deutsche Forschungsgemeinschaft (Sonderforschungsbereich 51 Medizinische Molekularbiologie und Biochemie, and Schwerpunktsprogramm Mechanismen toxischer Wirkungen von Fremdstoffen).

References

1. Sies, H., Gerstenecker, C., Menzel, H., Flohé, L.: FEBS Lett. 27, 171-175 (1972)
2. Sies, H., Gerstenecker, C., Summer, K.H., Menzel, H., Flohé, L.: In: Glutathione. Flohé, L., Benöhr, H., Sies, H., Waller, H.D., Wendel, A. (eds.), pp.261-275. Stuttgart: Thieme Verlag 1974
3. Sies, H., Summer, K.H., Grosskopf, M.: In: Use of isolated liver cells and kidney tubules in metabolic studies. Tager, J.M., Söling, H.D., Williamson, J.R. (eds.), pp.317-322. Amsterdam: North Holland 1975
4. Sies, H., Summer, K.H.: Eur. J. Biochem. 57, 503-512 (1975)
5. Oshino, N., Chance, B.: Biochem. J. 162, 509-525 (1977)
6. Nishiki, K., Jamieson, D., Oshino, N., Chance, B.: Biochem. J. 160, 343-355 (1976)
7. Burk, R.F., Nishiki, K., Lawrence, R.A., Chance, B.: J. Biol. Chem. 253, 43-46 (1978)

8. Sies, H., Bartoli, G.M., Burk, R.F., Waydhas, C.: Eur. J. Biochem. 89, 113-118 (1978)
9. Chance, B., Boveris, A., Nakase, Y., Sies, H.: This volume
10. Sies, H., Wahlländer, A., Linke, I., Marklstorfer, A.: Hoppe-Seyler's Z. Physiol. Chem. 359, 1151 (1978)
11. Bartoli, G.M., Sies, H.: FEBS Lett. 86, 89-91 (1978)
12. Sies, H.: Meth. Enzymol. 52, 48-59 (1978)
13. Racker, E.: J. Biol. Chem. 190, 685-696 (1951)
14. Tietze, F.: Anal. Biochem. 27, 502-522 (1969)
15. Habig, W.H., Pabst, M.J., Jakoby, W.B.: J. Biol. Chem. 249, 7130-7139 (1974)
16. Levine, W.G.: Ann. Rev. Pharmacol. Toxicol. 18, 81-96 (1978)
17. Chasseaud, L.F.: In: Glutathione. Flohé, L., Benöhr, H., Sies, H., Waller, H.D., Wendel, A. (eds.), pp.90-108. Stuttgart: Thieme Verlag 1974
18. Cohen, A.J., Smith, J.N., Turbert, H.: Biochem. J. 90, 457-464 (1964)
19. Jakoby, W.B.: Adv. Enzymol. 46, 383-414 (1978)

Reduction of Diazenes and Hydroperoxides by Rat Liver Mitochondria

P. C. JOCELYN

Summary

Rat liver mitochondria contain a complete glutathione redox system comprising GSH, glutathione reductase, glutathione peroxidase and $NADPH_2$ generated from $NADH_2$ by transhydrogenation. The capacity of this system has been tested by adding recognised thiol oxidants (diazenes and neutral hydroperoxides) to mitochondrial suspensions and measuring their rate of consumption together with the steady-state concentration of mitochondrial GSH. Diamide (a diazene) is reduced smoothly without added reducible substrates but reduction of hydroperoxides stops after an initial burst. With diamide, the mitochondrial GSH level remains near the initial value but with hydroperoxides it falls. The fall is prevented by adding succinate and hydroperoxide reduction then rises to a constant maximum rate of 14-20 nmol/min/mg protein. Reduction of all the oxidants is inhibited by N-ethylmaleimide, p-hydroxymercuribenzoate, malonate and antimycin but not by cyanide or oligomycin. Using diamide as the oxidant, these inhibitors correspondingly lower the mitochondrial GSH concentration demonstrating a close correlation between this parameter and the rate of oxidant reduction. Severe inhibition of oxidant reduction is obtained with uncouplers and other competitors for high energy intermediates. Succinate largely prevents the GSH levels from falling in the presence of an uncoupler and an oxidant but it does not relieve the inhibition of the reduction.

The results suggest that the glutathione redox system is chiefly responsible for the reduction of thiol oxidants and that both the penetration of the oxidant and the regeneration of GSH from GSSG are energy dependent processes.

A. Oxidants for the Glutathione Redox System

Rat liver mitochondria contain a small amount of glutathione (GSH), 4-8 nmol/mg protein, together with glutathione reductase (1) and glutathione peroxidase (1-3). They can also produce $NADPH_2$ via transhydrogenation from $NADH_2$ and so contain a complete redox system similar to that present in the cytosol. The properties and capacity of this system can only be found by allowing it to do work by adding various substances capable of oxidising GSH to GSSG and measuring the rate of their reduction under different conditions. Two types of thiol oxidant have been used, namely diazenes and hydroperoxides.

I. Diazenes

These substances, introduced by Kosower, have been extensively used as GSH oxidants in many different biological systems [for a review see (4)]. Though not specific for thiols, their rate of reaction with GSH is an order of magnitude greater than that obtained with $NAD(P)H_2$, flavins, ferredoxins, or dihydrolipoate (5).

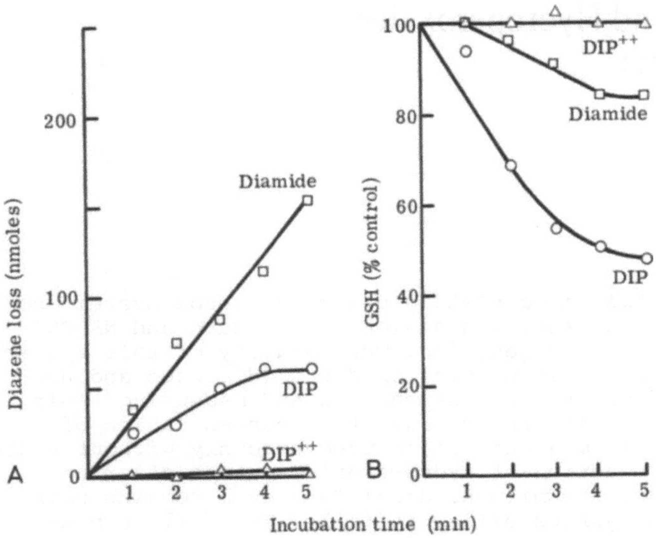

Fig. 1A and B. Comparison of the rate of mitochondrial reduction of diazenes (A) and the corresponding GSH concentrations (B). Mitochondria (3 mg protein containing 16.5 nmol GSH) were added to the diazene (0.5 µmol) in Tris-KCl buffer, pH 7.2 (1.4 ml). After incubation at 30°C for the stated time, the pellet was separated from the supernatant in an Eppendorf high-speed centrifuge. Diazene loss was determined from the loss of extinction at 310 nm and non protein thiols assayed with DTNB. The value for GSH is expressed as a percentage of the original amount remaining in the pellet

The four most used diazenes are as follows (trivial names given)

$(CH_3)_2N.CON{=}NCO.N(CH_3)_2$	Diamide
$C_6H_5 — N{=}NCO_2C_2H_5$	Azoester
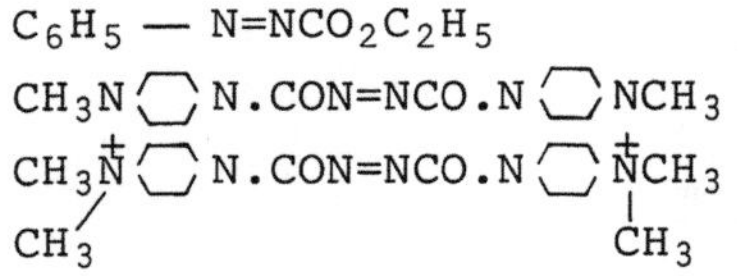	DIP
	DIP^{2+}

Azoester is unsuitable as an oxidant for mitochondrial GSH because, when added to mitochondria, it is itself oxidised (6). The other three diazenes behave as follows (Fig.1A). DIP^{2+} is unaffected, DIP undergoes a burst of reduction which then ceases, and diamide is reduced at a constant rate. To investigate the role of GSH in these reductions, the non protein thiol concentration in the sedimented pellet at intervals during the reduction has been determined. In the presence of DIP^{2+} the GSH content is unaffected and during the reduction with diamide it also remains at a high level. With DIP however, the value falls gradually to about a half of that found before incubation (Fig.1B). The effect of adding succinate on the reductions is shown in Figure 2. DIP^{2+} is still not reduced, whereas DIP is now reduced at a faster rate than diamide, whose rate of reduction remains largely unaffected. The fall in GSH with DIP (Fig.1B) is also largely prevented by succinate. The action of other Krebs cycle intermediates is shown in Figure 3. Except in the case of malate, there is good correlation between the effect of each acid on the rate of DIP reduction and on the GSH content. Thus, moderate increases in each parameter are obtained on adding citrate or

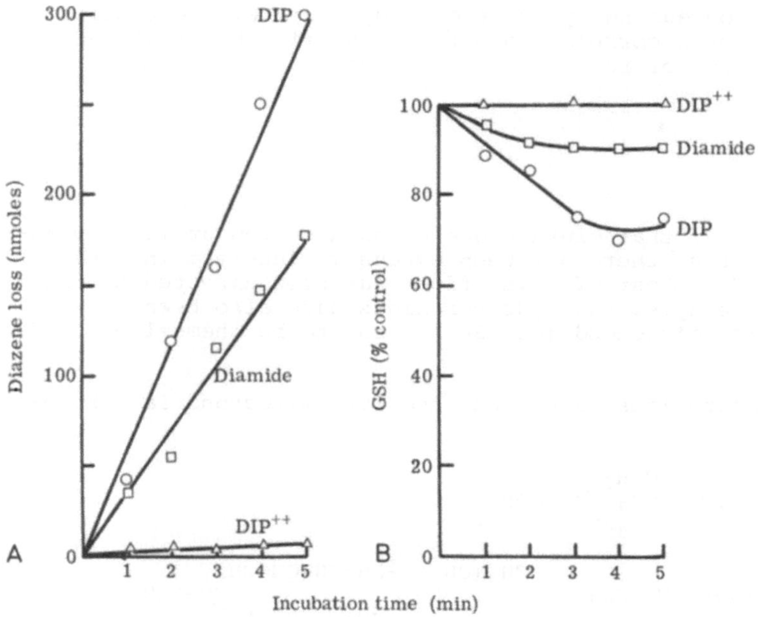

Fig. 2A and B. Effect of succinate (2.5 mM) on the rate of mitochondrial reduction of diazenes (A) and on the corresponding GSH concentration (B). Conditions as for Figure 1. Mitochondria were preincubated 2 min with succinate before adding the diazene

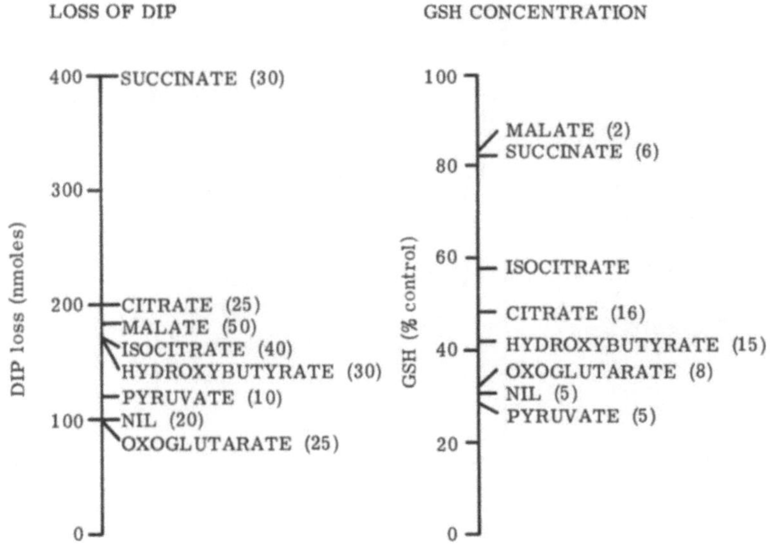

Fig. 3. Influence of Krebs cycle intermediates on the reduction of DIP and on the corresponding GSH concentration. Conditions as for Figure 2. Values were determined after 5 min incubation at 30°C. GSH is given as a percentage of the original amount found before incubation. Standard deviations in parentheses. Results from at least 3 experiments

isocitrate, but not on adding pyruvate or oxoglutarate. The results taken together suggest a correlation between the mitochondrial GSH concentration and the rate of reduction of diazene.

II. Hydroperoxides

Since the discovery (7) that glutathione peroxidase can utilise hydroperoxides as substrates, there has been extensive interest in their ability to oxidise GSH. Most of this effort has been directed to the role of the cytosol enzyme, but hydroperoxides have also been shown to be able to oxidise mitochondrial GSH (8) and to be themselves consumed (3).

The following hydroperoxides have been used with mitochondria (abbreviated names given)

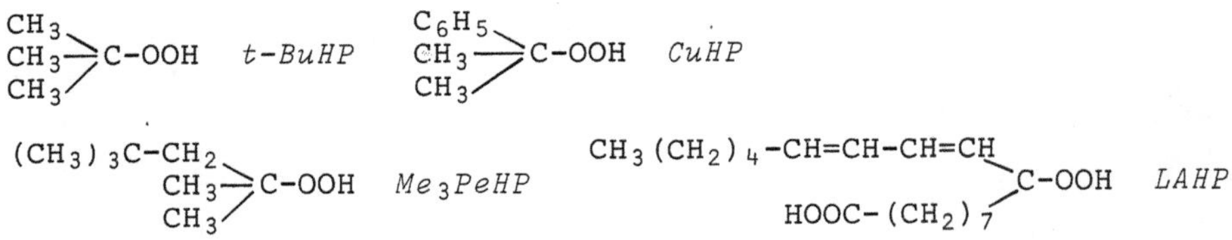

Their rates of reduction have been compared together with the corresponding mitochondrial GSH concentrations during the reaction.

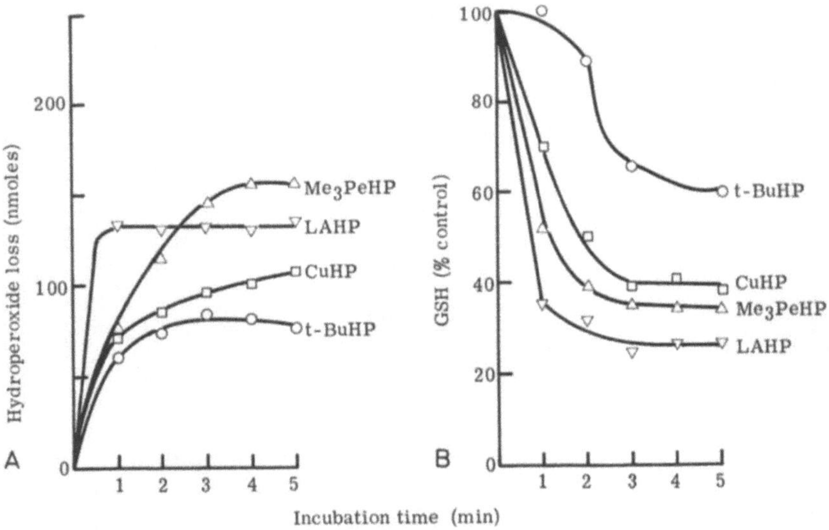

Fig. 4A and B. Comparison of the rate of reduction of some hydroperoxides (A) and the corresponding GSH concentrations (B). Conditions as for Figure 1 using 4 mg protein and 0.5 μmol hydroperoxide. Hydroperoxide concentrations during the incubation were determined after sedimenting the mitochondria by assaying the unacidified supernatant with added GSH and a glutathione peroxidase preparation

Initial rates of reduction without reducible substrates present are, in decreasing order, LAHP, Me_3PeHP, CuHP and t-BuHP (Fig.4A) but these rates fall to zero after 1-3 min, though much oxidant still remains. Except for t-BuHP, the corresponding concentrations of mitochondrial GSH fall in the same manner as the initial rates of reduction, the greatest fall being found with LAHP (Fig.4B). However, the correlation of GSH concentration with oxidant reduction does not apply to t-BuHP, where a high GSH is associated with a low rate of reduction. When succinate is added, the reduction rates of all the hydroperoxides except LAHP increase greatly to the same maximum value and remain constant until they are all consumed (Fig.5A). Corresponding GSH concentrations also remain high during the reaction (Fig.5B). The rate of reduction of LAHP and the corresponding low level of GSH are both unaffected by succinate, perhaps because LAHP penetrates much faster than the other hydroperoxides and overwhelms the reduction mechanism. The effect of other Krebs cycle intermediates on the reduction of t-BuHP is shown in Figure 6. The rate of reduction is greatest with succinate, but some other intermediates are also active, particularly hydroxybutyrate. As found with DIP, oxoglutarate and pyruvate do not support the reductions The corresponding GSH concentrations show some correlation with the rates of reduction of the hydroperoxides (i.e., a high reduction rate is accompanied by a high GSH level) except for malate, which with both hydroperoxides elevates the GSH concentration much more than the hydroperoxide reduction rate (as also found with DIP, see Fig.3).

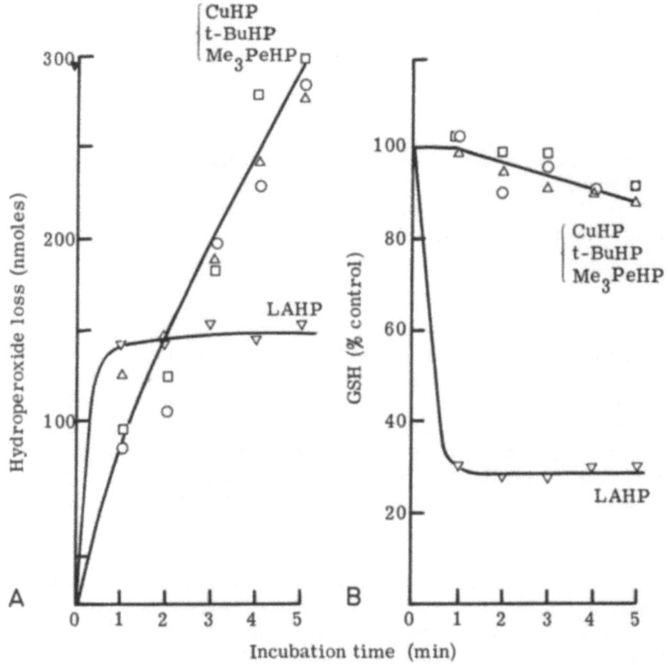

Fig. 5A and B. Effect of succinate (2.5 mM) on the rate of mitochondrial reduction of hydroperoxides (A) and on the corresponding GSH concentrations (B). Conditions as for Figure 4

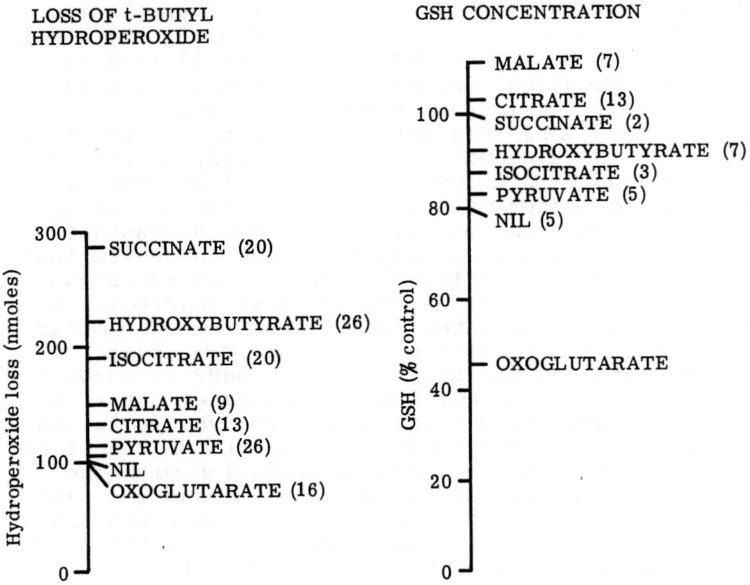

Fig. 6. Effect of Krebs cycle intermediates on the loss of t-butyl hydroperoxide (t-BuHP) and the corresponding GSH concentration. Conditions as for Figure 5. Values were determined after incubation for 5 min. GSH is given as a percentage of the original amount present before incubation. Means and differences are given for two assays

B. Inhibitors of Oxidant Reduction

Various substances inhibit the reductions of diazenes and hydroperoxides in the presence of mitochondria, and at the same time deplete the concentration of mitochondrial GSH. Their effects are exemplified in this section using diamide as the oxidant [see also (16)].

I. Thiol Agents

The best evidence that GSH is directly involved in diamide reduction comes from the effect of adding thiol agents (Fig.7). For instance, N-ethylmaleimide which is known to titrate mitochondrial GSH directly (9) is an effective inhibitor of diamide reduction, and when its concentration is varied there is a close correlation between the mitochondrial GSH concentration and the extent of inhibition. This correlation is also found with 2,4-dichlorphenol-indophenol, p-hydroxymercuriben-

Fig. 7. Effect of thiol agents at varying concentration on the reduction of diamide (x) and the corresponding GSH concentration (□). 1 ml Tris-KCl buffer containing mitochondria (~4 mg protein) and inhibitor at the concentration shown was preincubated at 30°C for 5 min. More buffer (0.3 ml) with or without diamide to give 0.35 mM was added and the mixture reincubated at 30°C for 5 min. GSH concentrations are given as a percentage of the amount present after incubation with diamide but without inhibitor added (see Fig.1B). Dotted lines show the GSH concentration in the same units when diamide is ommited. The fall in diamide is given as a percentage of the fall found without the addition of inhibitor

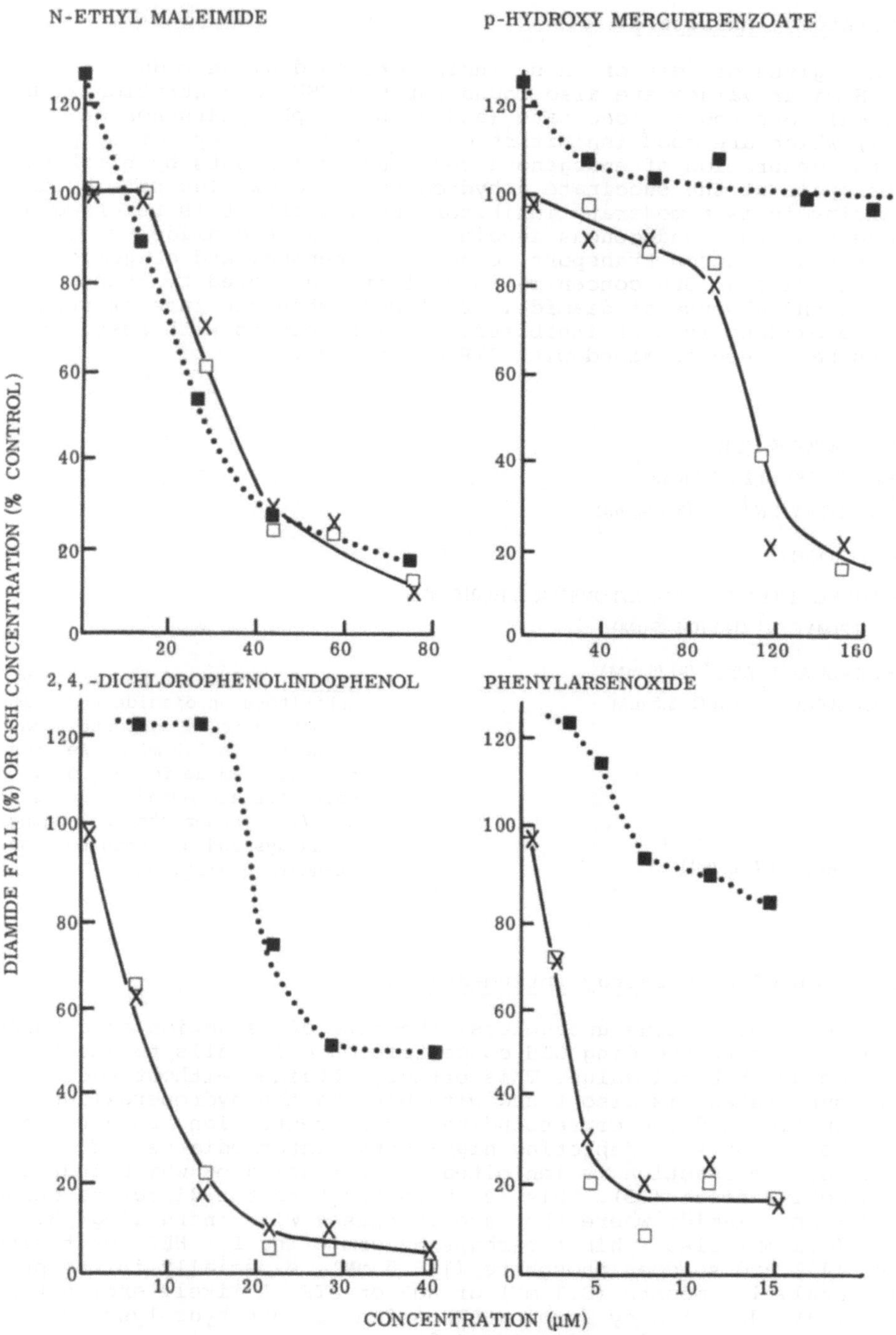

zoate and phenylarsenoxide, which inhibit at concentrations that do not appreciably affect the GSH concentration in the absence of oxidant. The effect of the mercurial is especially notable as it is known to be a non-penetrant (9). Such reagents presumably act on the GSH level by inhibiting some component of the GSH regeneration system. These thiol reagents also inhibit the reduction of t-BuHP and DIP.

II. Respiratory Inhibitors

Effects are given of some of these inhibitors on diamide reduction (Fig.8). Similar values are also found for the GSH concentration with diamide. Malonate and oxaloacetate (and probably phenylarsenoxide, see Fig.7) which are good inhibitors of the reduction may act by preventing the generation of endogenous reducing equivalents by blocking the Krebs cycle at the succinate dehydrogenase and oxoglutarate oxidase steps. Antimycin is a moderate inhibitor and its effect is reversed by ATP, suggesting that endogenous succinate can reduce diamide but only after reversed electron transport. Cyanide, rotenone, and oligomycin are poor inhibitors. GSH concentrations, little affected by these inhibitors in the absence of diamide, are lowest when the rate of reduction of the oxidant is most inhibited. Similar results with respiratory inhibitors have been obtained with DIP and t-BuHP.

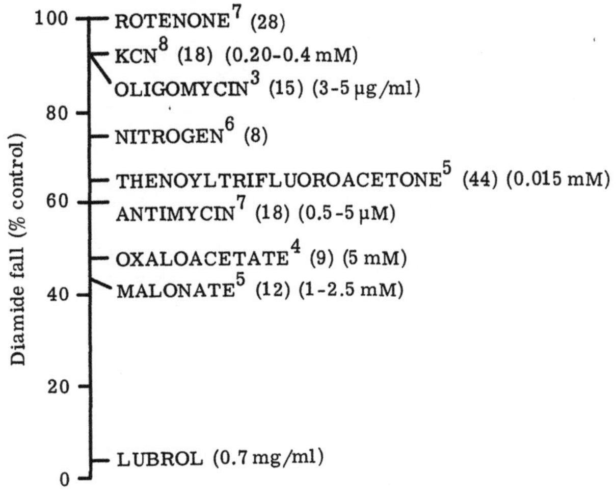

Fig. 8. Effect of respiratory inhibitors on diamide reduction. Mitochondria (4 mg protein) were incubated in 1.3 ml buffer containing diamide for 3-4 min. Units are expressed as in Figure 7. Superscripts show number of assays and brackets enclose standard deviations

III. Depletion of High-Energy Intermediates

In the presence of various uncouplers, the rate of reduction of diamide (Fig.9) and the corresponding GSH concentration each falls to about a third of the uninhibited value. This effect, obtained without added reducible substrates, is also found with DIP and the hydroperoxides. Diamide reduction and the corresponding GSH concentration are also inhibited by other ways of depleting high-energy intermediates (HEI). For instance, the reaction is inhibited by increasing pH which decreases the protonmotive force (10). This is in contrast to the direct reaction between GSH and diamide where the rate increases with increasing pH. Ca^{2+} ions (0.2 mM) also inhibit perhaps by competing for HEI for their own entry (11) and so does phosphate (1-2.5 mM), especially in the presence of catalytic amounts (0.1 mM) of ADP or ATP. A likely explanation is that HEI are depleted by forming ATP which is then hydrolysed by ATP-ase activity released by diamide (12).

This plethora of effects suggesting that high-energy intermediates may be consumed during diamide reduction indicates a need for their concurrent formation. This finding has to be reconciled with the lack of inhibition by cyanide (Sect.B.2). No evidence has been obtained that

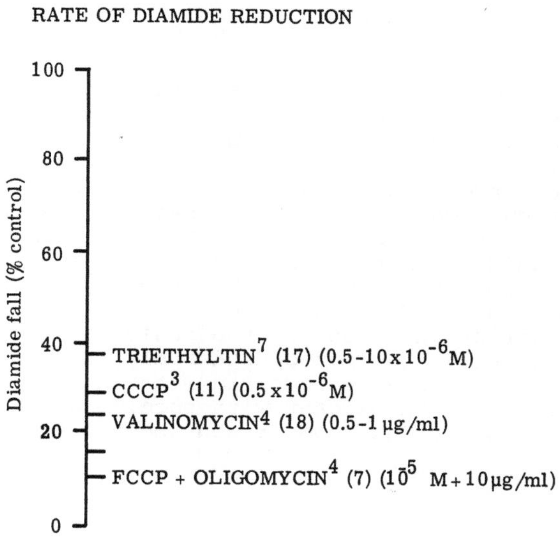

Fig. 9. Effect of uncouplers on diamide reduction. Units are expressed as in Figure 7. See also Figure 8

Table 1. Increments of ATP due to diamide. 1 ml Tris-KCl buffer containing 2 mg mitochondrial protein, 25 µg oligomycin and 0.3 mM cyanide were incubated with the additions shown and either with or without 0.35 mM diamide. After 2-4 min at 30°C, the proteins were precipitated with perchloric acid and ATP assayed with luciferase. The increments given were obtained by subtracting the ATP found when diamide was present from the corresponding value when diamide was absent. Means are given for three assays

Addition	ATP increment (nmol)
None	2.2 ± 1.0
N-ethylmaleimide (200 µM)	0.65 ± 0.25
Phenylarsenoxide (50 µM)	-1.4
Malonate (2.5 mM)	1.0 ± 0.1
Malonate plus isocitrate (2.5 mM)	7.5 ± 3.0
Malonate plus oxoglutarate (2.5 mM)	4.8 ± 2.0
Malonate plus succinate (2.5 mM)	-0.9 ± 0.9
Malonate plus hydroxybutyrate (2.5 mM)	-0.3 ± 0.3

diamide might itself also take electrons from the respiratory chain (i.e., sonic particles do not reduce it with e.g., succinate). In the presence of cyanide and oligomycin, endogenous ATP concentrations are higher when diamide is included in the incubation medium than when it is not, and this increment is inhibited by malonate, phenylarsenoxide, and N-ethylmaleimide. It is also much increased by some Krebs cycle intermediates especially isocitrate, citrate and oxoglutarate (Table 1). This extra ATP may be formed by flux through succinate thiokinase (13) induced by the $NADH_2$ consumption associated with diamide reduction. However, no positive ATP increment is obtained if oligomycin is omitted, and so this explanation may not be an adequate explanation for the origin of HEI.

C. The Permeability of Oxidants

Diazenes and hydroperoxides might be expected to penetrate into mitochondria by simple diffusion since they are soluble in non-polar solvents. However, with the exception of LAHP, they all appear to enter much more slowly after uncoupling. This is shown by the effect of adding a reducible substrate as well as an uncoupling agent (FCCP). Thus with, for example, isocitrate, diamide reduction is still inhibited but the GSH concentration now remains near to the original level (Fig.10). Similar findings are obtained with diamide and other Krebs cycle intermediates. The effect of succinate is also shown on the inhibition by FCCP of the reduction of all the oxidants used (Table 2, columns 2 and 3). In each case succinate has little effect in relieving the inhibition of reduction by the uncoupler previously noted but considerably increases the GSH concentration. The values may be compared with those previously given for reduction rates of each oxidant with succinate but without the uncoupler (Table 2, column 1). This comparison shows that

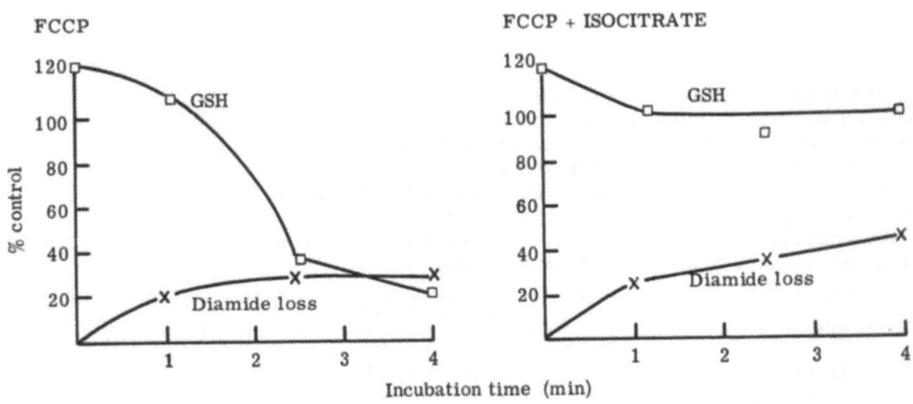

Fig. 10. Comparison of the effect of FCCP (0.5 μM) on diamide reduction and the corresponding GSH concentration with and without isocitrate present. Conditions and units as in Figure 9

Table 2. Effect of FCCP on oxidant reduction and GSH concentration with or without succinate present. Incubations (5 min at 30°C) were performed using 4 mg mitochondrial protein as described in Figures 3 and 5 with or without FCCP (0.5 μM) and succinate (2.5 mM). Means are given for at least two assays. Differences do not exceed 30 nmol (oxidant loss) or 13% (GSH content). GSH concentrations are given as a percentage of the amount present in the unincubated mitochondria

	With succinate		With FCCP		With FCCP and succinate	
Oxidant	Oxidant loss (nmol)	GSH content (%)	Oxidant loss (nmol)	GSH content (%)	Oxidant loss (nmol)	GSH content (%)
Diamide	210	90	60	37	70	90
DIP	398	82	60	35	85	65
t-BuHP	290	100	55	40	54	88
CuHP	271	88	63	35	95	74
Me_3PeHP	275	92	90	32	115	54

the uncoupler lowers the reduction rates drastically to basal values which differ somewhat for each oxidant. In contrast, the values for GSH obtained with succinate are not much altered as a result of adding the uncoupler (a significant decline is found with Me_3PeHP which has the highest reduction rate with uncoupler).

D. Mechanism of Oxidant Reduction

The pathway for oxidant reduction by mitochondria outlined in Figure 11 will be considered in relation to the evidence presented.

Reduction of diazenes and hydroperoxides occurs ultimately from reducing equivalents (endogenous or added) supplied by Krebs cycle acids. Of these, pyruvate and oxoglutarate are ineffective possibly because the oxidants convert the coenzyme A required for their dehydrogenation to the inactive disulphide form. Thus the oxidants also prevent oxygen uptake from these acids (3).

$NADPH_2$ required for glutathione reductase is generated via transhydrogenation and, in the case of succinate, reversed electron transfer. Electrons are then passed to regenerate GSH from GSSG formed by reduction of the oxidant. In the process, the pool of reduced nicotinamide nucleotides tends to fall but, as shown with a pulse of t-butyl HP (14,15), this is prevented by the uncoupler and antimycin sensitive action of succinate.

Much evidence supports the conclusion that GSH is the proximate reductant for the oxidants. With diamide, the most studied oxidant, the rate of reduction is closely related to the mitochondrial GSH concentration determined during the reduction in the presence of various inhibitors.

The GSH content of mitochondria is small (4-8 nmol/mg protein) compared to the maximum rates of oxidant reduction (~ 15 nmol/mg protein/min) and it must therefore be continually regenerated. Its steady-state concentration during the reduction thus depends on three processes:

1. the rate of regeneration from GSSG,
2. the rate of reaction with the oxidant,
3. the rate of penetration of the oxidant.

In coupled mitochondria without added substrates, (3) limits the rate of reduction of diamide and perhaps t-BuHP because the GSH concentration remains high and (1) the rate of reduction of DIP, LAHP, CuHP, and Me_3PeHP because the GSH falls during the reactions. With succinate present (3) still limits the rate with diamide and perhaps DIP and (1) still limits the rate with LAHP (both rates are unaffected by the succinate) but for the other oxidants the rate is limited now by (2) [(3) would not be compatible with the fact that with succinate, reduction rates are approximately equal, whereas rates of penetration are not likely to be identical]. (2) may depend on the mitochondrial glutathione peroxidase activity.

$$\text{SUBSTRATES} \rightarrow \text{NAD} \xrightarrow[\sim]{\text{transhydrogenase}} \text{NADP} \xrightarrow[\text{reductase}]{\text{glutathione}} \text{GSSG} \underset{\sim}{\rightarrow} \text{THIOL OXIDANT}$$

Fig. 11. Proposed route for oxidant reduction ~ represents the utilisation of high energy intermediates (HEI)

When mitochondria are uncoupled, the picture changes because (with the exception of LAHP, whose rate is already low) the dominant finding is a large decline in the rate of reduction of each of the oxidants which is little affected by adding reducible substrates. The decline is due to a big fall in (3) after uncoupling. A possible explanation is that protonated forms of the oxidants (e.g., RCON=$\overset{+}{N}$HCOR' and ROO$\overset{+}{H}_2$) can enter coupled mitochondria faster than the neutral forms because of the favourable membrane potential. After uncoupling, GSH levels with oxidant present but without added substrates fall because, by the proposed pathway, both (1) and (3) depend on the availability of HEI and (1) is most affected by their depletion. Because of the smaller oxidative stress, substrates such as succinate can increase (1) by supplying electrons faster than endogenous substrates and so shifting the redox balance to the right.

E. Conclusions

The maximum rate of reduction of hydroperoxides by isolated mitochondria corresponds to a mitochondrial contribution in intact tissue at 30°C of ~ 0.5 µmol/min/g liver. Perfused liver removes 3-4 µmol/min/g liver at 37°C and this is decreased by about 50% on lowering the temperature to 30°C (15). The inference, that mitochondria may account for ~25% of the total hydroperoxide reduction rate, compares with another estimate, based on different sensitivities in the loss of reduced nicotinamide nucleotides from hepatocyte suspensions, of 40% (14). The mitochondrial contribution may, however, be strongly dependent on the metabolic state of the cell. In state 4 the rate is maximal and available to the whole cell but in state 3, $NADPH_2$ production via energy-dependent transhydrogenation decreases, and so does the rate of entry of oxidant into the mitochondria. The effect is to allow mitochondria, insulated from external oxidant stress, to conserve their reducing equivalents for use in the regeneration of cellular supplies of ATP.

References

1. Flohé, L., Schlegel, W.: Hoppe-Seyler's Z. Physiol. Chem. 352, 1401-1410 (1971)
2. Neubert, D., Wojtczak, A.B., Lehninger, A.L.: Proc. Natl. Acad. Sci. U.S.A. 48, 1651-1658 (1962)
3. Sies, H., Moss, K.M.: Eur. J. Biochem. 84, 377-383 (1978)
4. Kosower, E.M., Kosower, N.S.: Nature (London) 224, 117-128 (1969)
5. O'Brien, R.W., Weitzman, P.D.J., Morris, J.G.: FEBS Lett. 10, 343-345 (1970)
6. Jocelyn, P.C.: Biochem. Pharmac. 25, 1267-1270 (1976)
7. Little, C., O'Brien, P.J.: Biochem. Biophys. Res. Commun. 31, 145-150 (1968)
8. Jocelyn, P.C.: Biochim. Biophys. Acta 369, 427-436 (1975)
9. Gaudemer, Y., Latruffe, N.: FEBS Lett. 54, 30-33 (1975)
10. Mitchell, P.: Biochem. Soc. Trans. 4, 399-430 (1976)
11. Brand, M.D., Lehninger, A.L.: J. Biol. Chem. 250, 7958-7960 (1975)
12. Siliprandi, D., Scutari, G., Zoccarato, F., Siliprandi, N.: FEBS Lett. 42, 197 (1974)
13. Lowenstein, J.M.: Comprehensive Biochemistry. Florkin, M., Stotz, E.H. (eds.), 18S, p.260. Amsterdam: Elsevier 1971
14. Oshino, N., Chance, B.: Biochem. J. 162, 509-525 (1977)
15. Sies, H., Summer, K.H.: Eur. J. Biochem. 57, 513-520 (1975)
16. Jocelyn, P.C.: Biochem. J. 176, 649-664 (1978)

Effect of Oxidation of Glutathione and Membrane Thiol Groups on Mitochondrial Functions

N. SILIPRANDI, D. SILIPRANDI, A. BINDOLI, and A. TONINELLO

Summary

Diamide [diazenedicarboxylic acid bis (N,N'-dimethylamide)] a thiol oxidant, does not titrate more than 25% of total liver mitochondrial thiol groups. These include the nonprotein thiols present in the intermembrane and matrix spaces and vicinal pairs of protein membrane-bound -SH. Mitochondrial glutathione content is decreased by diamide only at concentrations above 0.5 mM. At concentrations below those required for GSH oxidation, diamide slightly stimulates ATPase activity of intact rat liver mitochondria and induces an efflux of endogenous Mg^{2+} dependent on coupled respiration. These effects are fully preventy by EGTA and ruthenium red, a known inhibitor of Ca^{2+} transport across mitochondrial inner membrane. The dependence of these phenomena on an energy-dissipating Ca^{2+} flux is also supported by a release of state 4 respiration. In the presence of low concentrations of diamide (< 0.25 mM) external Ca^{2+} are rapidly taken up and retained within mitochondria, while Mg^{2+} are released in an energy-dependent process. It is assumed that diamide-induced Mg^{2+} efflux is the consequence of a cyclic in and out movement of Ca^{2+}, in which the passive efflux promoted by the oxidation of pairs of membrane-bound thiols is compensated by a continuous energy-dependent reuptake. At higher concentrations (> 0.5 mM) Ca^{2+} are initially taken up by liver mitochondria but subsequently released. Concomitantly the efflux of Mg^{2+}, initially dependent on respiration and sensitive to ruthenium red, becomes "passive" and insensitive to ruthenium red in coincidence with Ca^{2+} release. It is assumed that the redox state of some pairs of mitochondrial -SH (those sensitive to low diamide concentrations) might control, through a modulation of Ca^{2+} flux, the binding of Mg^{2+} to mitochondrial membranes as well as their permeability properties.

A. Introduction

The role of glutathione and related enzymes (glutathione peroxidase and reductase) in mitochondria is poorly known. As suggested by Vignais and Vignais (1973), glutathione may be a reservoir of reducing equivalents capable of buffering the effects of oxidants on -SH groups of protein and nonprotein thiols present in the inner mitochondrial compartment. Whether the glutathione system might also control inner mitochondrial membrane permeability by modifying the redox state of protein-bound -SH, is a matter of speculation. Up to now most of the work has been devoted to the study of glutathione-related enzymes (Flohé and Schlegel, 1971) and to the effects of thiol-reacting agents

---

Abbreviations: Diamide = diazenedicarboxylic acid bis (N,N-dimethylamide); DTE = dithioerythritol; DTNB = 5,5'-dithio-bis-(2-nitrobenzoic acid); EGTA = ethylene glycol-bis-(2-aminoethyl-ether)-N,N'-tetracetic acid; GSH = reduced glutathione; GSSG = oxidized glutathione; pCMB = para-chloromercuribenzoate; SDS = sodium dodecyl sulfate.

on glutathione content in mitochondria (Riley and Lehninger, 1964; Vignais and Vignais, 1973; Jocelyn and Kamminga, 1974; Gaudemer and Latruffe, 1975; Jocelyn, 1975; Bindoli et al., 1976).

Glutathione, as well as other nonprotein thiols, does not appreciably leak from the matrix space of intact mitochondria and remains stable during short-term incubation (Jocelyn, 1975). Under the usual incubation conditions no significant oxidation of these thiols takes place, unless thiol reagents are added (Jocelyn, 1975). As reported by Sies and Moss (1978) the oxidation of mitochondrial GSH by hydroperoxides is mediated by glutathione peroxidase. The action of cystamine (Sies and Moss, 1978), fuscin (Vignais and Vignais, 1973) and diamide (Siliprandi et al., 1974a) are thought to be independent from glutathione peroxidase and consequently less specific. Furthermore, at concentrations below those required to affect mitochondrial GSH, diamide induces a number of effects, which are attributable to its action on mitochondrial membranes (Siliprandi et al., 1974b, 1975a, 1975b, 1975c, 1977, 1978; Zoccarato et al., 1977). In the present paper we discuss the effect of diamide on total thiol and glutathione content of rat liver mitochondria in relation with some energy-linked functions, such as ATPase activity, mitochondrial swelling and cation flux across inner mitochondrial membrane.

B. Effect of Diamide on Total -SH and Glutathione Content of Rat Liver Mitochondria

On the ground of the reaction mechanism proposed for diamide (Kosower et al., 1972) beside glutathione and other nonprotein thiols present in the matrix and intermembrane spaces, it can be predicted that also unhindered pairs of protein thiol groups react with diamide. In order to know whether and to what extent mitochondrial -SH are affected by diamide, rat liver mitochondria have been titrated with diamide and pCMB. Figure 1 shows that 1 mM pCMB titrates almost all the mitochondrial -SH (94 ± 5 nmol/mg protein). This figure nicely agrees with that reported by Lehninger using amperometric titration (Riley and Lehninger, 1964). On the contrary 1 mM diamide does not titrate more than 25% of the total thiols. Since in solution at pH 7 diamide reacts very rapidly with glutathione, the requirement of a long incubation period (10 min) in order to attain the maximum effect, is, in all probability, expression of the slow penetration of diamide across mitochondrial membrane. Higher concentrations of diamide, or longer incubation times, do not modify the amount of oxidized -SH. The maximum amount of -SH titrated by diamide (approximately 23 nmol/mg protein) accounts for the sum of mitochondrial soluble thiols and pairs of membrane-bound -SH. Since the total amount of mitochondrial nonprotein -SH does not exceed 8-9 nmol/mg protein [~ 5 nmol of glutathione (Fig.2) and ~2 nmol of CoA (Bremer et al., 1972)], most of the -SH titrated by diamide should be protein -SH.

The total amount of protein -SH titrated by diamide is lower in sonic particles than in intact mitochondria (results not reported). This observation supports the proposal, implicit in diamide action mechanism (Kosower et al., 1972), that protein -SH titrated by diamide should be paired and sterically vicinal. It is conceivable that disruption of mitochondria decreases the number of vicinal -SH susceptible to react with diamide, by altering the arrangement of membrane-bound protein -SH.

The specific action of diamide on glutathione content of rat liver mitochondria is shown in Figure 2. It can be seen that after 10 min incuba-

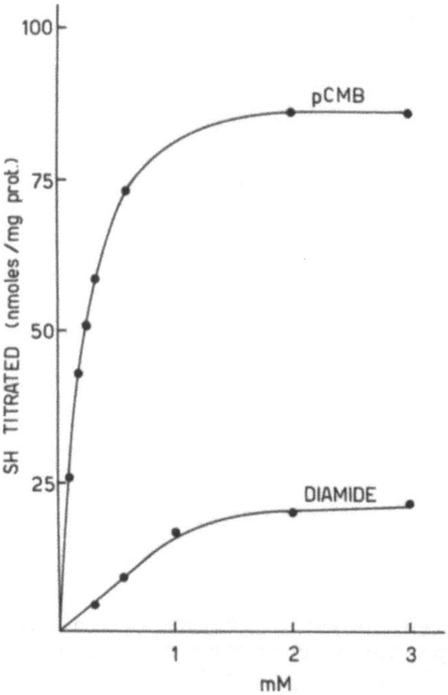

Fig. 1. Effect of diamide and pCMB on the amount of total thiol groups of rat liver mitochondria.
Rat liver mitochondria were isolated in 0.25 M sucrose buffered with 3 mM Tris-HCl at pH 7.4 by the method of Myers and Slater (1957). The protein content was estimated by the biuret method (Gornall et al., 1949). Incubation of rat liver mitochondria (2.5 mg/ml) was carried out at 25°C for 10 min in 0.125 M KCl, 20 mM Tris-HCl buffer pH 7.4 in the presence of various amounts of diamide or pCMB. At the end of incubation mitochondria were rapidly centrifuged, washed, and resuspended in the same medium. The unreacted protein -SH groups of mitochondria were titrated with DTNB (Ellman, 1959) in 2 ml of a mixture containing 0.2 M Tris-HCl buffer pH 8.1, 5 mM EDTA, 0.3% SDS, 1 mM DTNB and about 1 mg protein. Optical density was followed at 412 nm until the reaction was completed

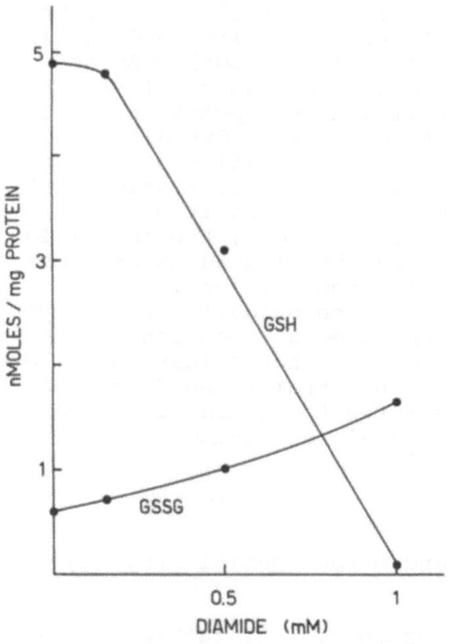

Fig. 2. Oxidation of mitochondrial glutathione by diamide.
Rat liver mitochondria were incubated as described in Figure 1. Reduced and oxidized glutathione were estimated by the method of Tietze (1969)

tion in the absence of oxidizable substrates, the concentration of diamide which affects mitochondrial GSH is of the order of 0.5 mM. At 1 mM diamide all GSH disappears. Figure 2 also shows that the level of GSSG does not account for the amount of GSH which has disappeared.

Table 1. Effect of diamide on the oxidation of pyruvate and acetylcarnitine by rat kidney mitochondria

Diamide concent. mM	Pyruvate	Acetylcarnitine
	natoms O/mg protein	
0	60	76.8
0.05	60	58
0.1	58	43.2
0.25	54.8	40.8
0.5	48	16
1	38	8

Rat kidney mitochondria (5 mg protein) prepared in 0.25M sucrose + 1 mM EDTA were suspensed in a medium containing 0.125 KCl, 10 mM Tris-HCl pH 7.4, 1 mM Pi. Final volume 2 ml. Temperature 25°C. The substrates (5 mM pyruvate or 1 mM acetylcarnitine plus 0.1 mM malate) and 1 mM ADP were added after 5 min preincubation in the presence of diamide. The results are relative to oxygen consumption during the first 3 min after substrate addition.

The reason for this discrepancy is not known, however it can be in part explained by the formation of mixed disulphides with protein -SH, as demonstrated by Jocelyn and Kamminga (1974), and in part by a leakage of GSSG into the external medium.

A depletion of GSH has also been observed upon incubation of rat liver mitochondria in the presence of phosphate (Bindoli et al., 1976). However, under this condition the species released from mitochondria was found to be reduced glutathione, thus indicating that both GSSG and GSH can leak out of the mitochondria. Hence the problem still remains open as to whether diamide oxidizes GSH following the penetration into the inner mitochondrial compartment, or it reacts with GSH diffused into the external medium as a consequence of an alteration of mitochondrial permeability. In this connection it is relevant to mention that diamide inhibits pyruvate oxidation at the same concentration which affects mitochondrial GSH content (Table 1). Therefore, the inhibition of pyruvate (or α-ketoglutarate) oxidation can be assumed as an index, alternative to GSH disappearance, for diamide penetration into the inner mitochondrial compartment. Unexpectedly the oxidation of acetyl carnitine by rat kidney mitochondria is inhibited at much lower concentrations of diamide than pyruvate (Table 1). Since oxidation of both pyruvate and acetyl carnitine is dependent on intramitochondrial CoA, the different sensitivity of the two substrates to diamide is probably due to an inhibition of acetyl carnitine translocase (Pande and Parvin, 1976).

C. Action of Diamide on ATPase Activity and Mitochondrial Swelling

At concentrations which do not affect mitochondrial glutathione, diamide stimulates ATPase activity of intact liver mitochondria in a concentration-dependent manner (Table 2). Thus, ATPase stimulation does not appear to be the result of diamide penetration into the inner compartment, but rather the consequence of an inner membrane disarrangement induced by the reagent.

Table 2. Effect of diamide on ATPase activity of rat liver mitochondria

Diamide concentration mM	nmoles Pi x mg $prot^{-1}$ x min^{-1} hydrolyzed
0	13.7
0.075	14
0.15	27.2
0.3	37
0.4	47.6
0.5	58

5 mg of mitochondrial protein were added to a medium containing 80 mM KCl, 60 mM sucrose, 1 mM ATP and 5 mM Tris. HCl pH 7.4. Final volume 4 ml. Temperature 25°C. ATPase activity was estimated from Pi released (Fiske and Subbarow, 1925).

Table 3. Action of different agents on diamide stimulated ATPase and oxidation of mitochondrial glutathione

Additions	Phosphate hydrolyzed nmoles/mg prot/min	GSH nmoles/mg prot
None	18	5.25
Diamide 0.5 mM	60	2.85
" + succinate 1 mM	33	4.85
" + EGTA 1 mM	32	5.15
" + ruthenium red 5 µM	38	4.60

Experimental conditions as described in Figure 2 and Table 2.

Succinate prevents both the stimulation of ATPase activity and GSH decrease induced by 0.5 mM diamide (Table 3). Similar results are reported by Jocelyn (1975), and interpreted as due to a reduction of GSSG by NAD(P)H via glutathione reductase. Such an interpretation does not explain the inhibition of diamide-stimulated ATPase activity by succinate. It appears more likely that in the presence of a readily oxidizable substrate, the inner membrane undergoes conformational changes which render target -SH less accessible to diamide. Furthermore EGTA and ruthenium red, the well-known inhibitor of calcium transport across inner mitochondrial membrane, also prevent the action of diamide on ATPase activity and mitochondrial GSH content. It would then appear that at least some diamide effects might be mediated by a mobilization of endogenous Ca^{2+}. When the transport of this cation is inhibited by ruthenium red, or when Ca^{2+} released from mitochondria is trapped by EGTA, also diamide action is prevented.

This interpretation is also supported by the close similarity of rat liver mitochondrial swelling induced by diamide and by Ca^{2+} (Fig.3). In both cases the swelling, initiated by 1 mM phosphate, was inhibited both by EGTA and ruthenium red. Therefore, it is very likely that in both cases mitochondrial swelling reflects the same alterations of inner membrane induced either by added Ca^{2+} or by endogenous Ca^{2+} mobilized by diamide.

This conclusion is also supported by the experiments reported in Figure 4 which shows the additive effect of Ca^{2+} and diamide and the antagonism between Ca^{2+} and DTE. In the presence of Ca^{2+} plus diamide the swelling triggered by phosphate is more pronounced than that following the single addition of Ca^{2+} or diamide. Conversely, when Pi is

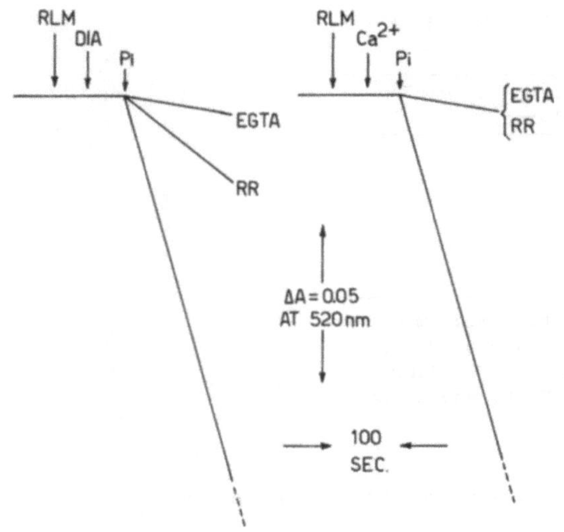

Fig. 3. Inhibition of Ca^{2+} and diamide-induced swelling by ruthenium red and EGTA.
Rat liver mitochondria (1 mg/protein) were suspended in a medium (final volume 2 ml) containing: 250 mM sucrose, 10 mM Tris-HCl pH 7.4, 5 mM potassium succinate, 2.5 μM rotenone. 0.5 mM Pi was added. Temperature 25°C. When present: 0.1 mM diamide, 0.1 mM $CaCl_2$, 0.1 mM EGTA, 2.5 μM ruthenium red. Swelling was monitored by absorbance at 520 nm using an Aminco-Chance spectrophotometer

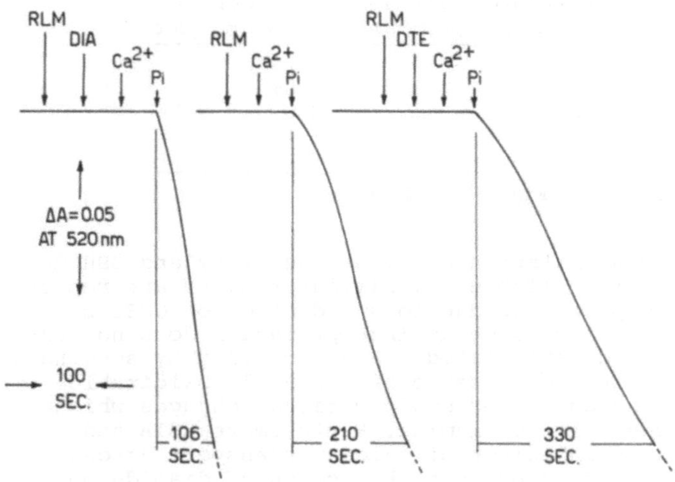

Fig. 4. Synergistic action of diamide and Ca^{2+} in promoting rat liver mitochondrial swelling.
Experimental conditions as in Figure 3. When present 1 mM DTE

added to mitochondria incubated in the presence of Ca^{2+} plus DTE, the swelling was markedly reduced. It would then appear that Ca^{2+} are in some way prevented by some membrane thiol groups from reaching the sites responsible for the maintenance of mitochondrial permeability. When these are oxidized by diamide, Ca^{2+} action is facilitated, whereas when they are protected by a proper reductant (DTE), Ca^{2+} action is prevented.

D. Action of Diamide on Mitochondrial Ca^{2+} and Mg^{2+} Flux

The antagonistic effects of Ca^{2+} and Mg^{2+} on diamide-induced mitochondrial swelling (Siliprandi et al., 1975b) prompted us to study the movements of these cations in diamide-treated mitochondria.

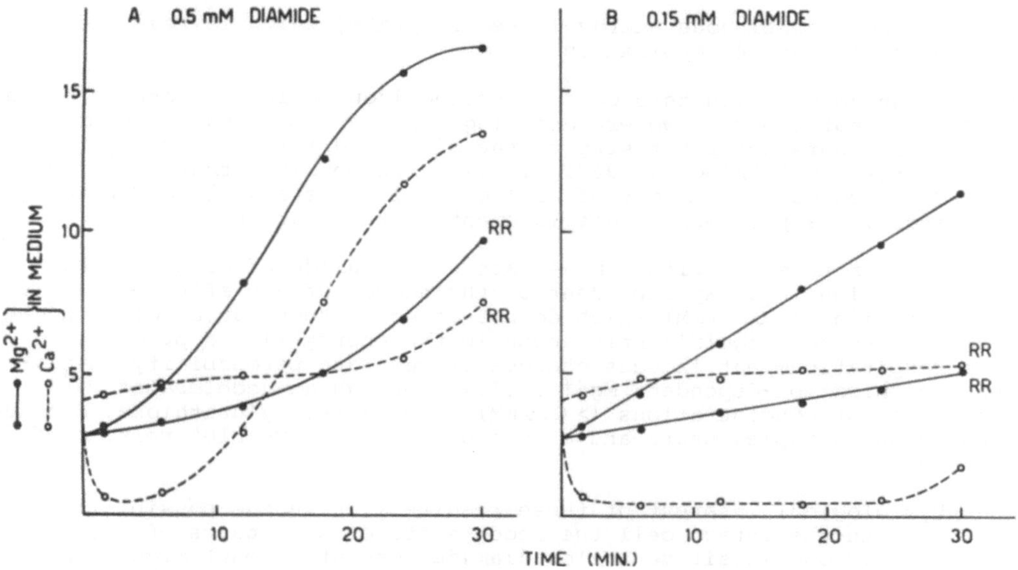

Fig. 5A and B. Flux of Ca^{2+} and Mg^{2+} in rat liver mitochondria. Effect of two different concentrations of diamide.
Rat liver mitochondria were suspended (1 mg/ml) in a medium containing 170 mM sucrose, 10 mM Tris-HCl pH 7.4, 5 mM succinate, 1.25 μM rotenone. When present 5 μM ruthenium red. Temperature 25°C. Mg^{2+} and Ca^{2+} movements were estimated by atomic absorption spectroscopy on the supernatant according to Crompton et al. (1976). Cation contents at zero time were in nmol/mg protein: Ca^{2+} 18; Mg^{2+} 22

The results of Figure 5A were obtained with 0.5 mM diamide, a concentration capable of decreasing mitochondrial glutathione and inducing an uncoupled state in about 8 min. All the Ca^{2+} present in the incubation medium as contaminant is taken up rapidly, but after about 8 min, when the respiratory control ratio is decreased to a value around 2, a progressively increasing Ca^{2+} efflux occurs. No Ca^{2+} efflux takes place in the absence of diamide. Concomitantly a respiration-dependent efflux of Mg^{2+} does occur. Mg^{2+} efflux is completely prevented by EGTA, whereas ruthenium red inhibits the initial efflux, but not the subsequent one concomitant with Ca^{2+} release. Evidently diamide, at the moment in which it reaches simultaneously mitochondria lose the capability to reaccumulate Ca^{2+}. As soon as this condition is attained ruthenium red becomes unable to further inhibit Ca^{2+} efflux. Since uncouplers do not induce any Mg^{2+} efflux (Siliprandi et al., 1978), these effects are not explainable by the uncoupled state induced by high concentrations of diamide.

Figure 5B reports the time course of Ca^{2+} and Mg^{2+} movements induced by 0.15 mM diamide, a concentration below that required to induce an uncoupled state as well as a decrease of mitochondrial GSH. Under this condition Ca^{2+} is retained within mitochondria, whereas an efflux of Mg^{2+}, linear with the time, occurs. Ruthenium red fully prevents the efflux of both Ca^{2+} and Mg^{2+}.

It is interesting to note that analogous fluxes of Ca^{2+} and Mg^{2+} have been induced with ionophore A23187 (Reed and Lardy, 1972). It is conceivable that diamide increases the passive efflux of accumulated Ca^{2+} by oxidizing some pairs of membrane-bound -SH. Such an efflux is compensated by a continuous energy-linked reuptake, which is revealed by a release of state 4 respiration.

The mechanism by which this Ca^{2+} recycling induces in its turn an efflux of Mg^{2+} is not clear. However, assuming that Ca^{2+} binding to inner membrane represents the first step of the mitochondrial Ca^{2+} transport (Reynafarje and Lehninger, 1969), it is quite possible that a continuous inward and outward transport of Ca^{2+} might interfere with Mg^{2+} binding, thus inducing a progressive displacement of this cation.

All together these results demonstrate that diamide affects mitochondrial function in a way dependent on the number of -SH affected. At low concentrations (< 0.25 mM) which do not affect mitochondrial glutathione, diamide does not significantly uncouple the energy-linked processes from respiration, but induces changes of membrane permeability involving a respiration-dependent Mg^{2+} efflux, through an accelerated Ca^{2+} recycling. At concentrations (> 0.5 mM) which affect glutathione, diamide induces an uncoupled state and a consequent "irreversible" release of cations.

The physiological interest of these results lies in the possibility that also in the intact cell the redox state of some pairs of mitochondrial -SH (those sensitive to low diamide concentrations) might control Mg^{2+} binding to the membrane and the flux of other cations through a modulation of the rate of Ca^{2+} inward and outward flux.

References

Bindoli, A., Cavallini, L., Siliprandi, N., Zoccarato, F.: Action of some thiol oxidizing reagents on mitochondrial sulfhydryl groups. Bull. Mol. Biol. Med. 1, 92-96 (1976)

Bremer, J., Wojtczak, A., Skrede, S.: The leakage and destruction of CoA in isolated mitochondria. Eur. J. Biochem. 25, 190-197 (1972)

Crompton, M., Capano, M., Carafoli, E.: Respiration-dependent efflux of magnesium ions from heart mitochondria. Biochem. J. 154, 735-742 (1976)

Ellman, G.L.: Tissue sulfhydryl groups. Arch. Biochem. Biophys. 82, 70-77 (1959)

Fiske, C.H., Subbarow, Y.: The colorimetric determination of phosphorus. J. Biol. Chem. 66, 375-400 (1925)

Flohê, L., Schlegel, W.: Glutathion-Peroxidase. IV. Intrazelluläre Verteilung des Glutathion-Peroxidase Systems in der Rattenleber. Hoppe-Seyler's Z. Physiol. Chem. 352, 1401-1410 (1971)

Gaudemer, Y., Latruffe, N.: Evidence for penetrant and non-penetrant thiol reagents and their use in the location of rat liver mitochondrial D(-)-β-hydroxybutyrate dehydrogenase. FEBS Lett. 54, 30-34 (1975)

Gornall, A.G., Bardawill, C.J., David, M.M.: Determination of serum proteins by means of the biuret reaction. J. Biol. Chem. 177, 751-766 (1949)

Jocelyn,P.C.: Some properties of mitochondrial glutathione. Biochim. Biophys. Acta 396, 427-436 (1975)

Jocelyn, P.C., Kamminga, A.: The non-protein thiol of rat liver mitochondria. Biochim. Biophys. Acta 343, 356-362 (1974)

Kosower, E.M., Correa, W., Kinon, B.J., Kosower, N.S.: Glutathione VII. Differentiation among substrates by the thiol-oxidizing agent, diamide. Biochim. Biophys. Acta 264, 39-44 (1972)

Myers, D.K., Slater, E.C.: The enzymic hydrolysis of adenosine triphosphate by liver mitochondria I. Activities at different pH values. Biochem. J. 67, 558-572 (1957)

Pande, S., Parvin, R.: Characterization of carnitine acylcarnitine translocase of heart mitochondria. J. Biol. Chem. 251, 6683-6691 (1976)

Reed, P.W., Lardy, H.A.: A23187: a divalent cation ionophore. J. Biol. Chem. 247, 6970-6977 (1972)

Reynafarje, B., Lehninger, A.L.: High affinity and low affinity binding of Ca^{++} by rat liver mitochondria. J. Biol. Chem. 244, 584-593 (1969)

Riley, M.V., Lehninger, A.L.: Changes in sulfhydryl groups of rat liver mitochondria during swelling and contraction. J. Biol. Chem. 239, 2083-2089 (1964)

Scutari, G., Siliprandi, D., Zoccarato, F.: Action of diamide on some energy linked processes of rat liver mitochondria. FEBS Lett. 42, 197-199 (1974a)

Sies, H., Moss, K.M.: A role of mitochondrial glutathione peroxidase in modulating mitochondrial oxidations in liver. Eur. J. Biochem. 84, 377-383 (1978)

Siliprandi, D., Scutari, G., Zoccarato, F., Siliprandi, N.: Sulfhydryl groups oxidation and mitchondrial functions. In: Membrane proteins in transport and phosphorylation. Azzone, G.F., Klingenberg, M.E., Quagliariello, E., Siliprandi, N. (eds.), pp.265-274. Amsterdam: North Holland Publ. Co. 1974b

Siliprandi, N., Siliprandi, D., Bindoli, A., Rugolo, M., Toninello, A., Zoccarato,F.: Action of diamide on ATPase activity of rat liver mitochondria. In: Electron transfer chains and oxidative phosphorylation. Quagliariello, E., Papa, S., Palmieri, F., Slater, E.C., Siliprandi, N. (eds.), pp.439-444. Amsterdam: North-Holland Publ. Co. 1975c

Siliprandi, N., Siliprandi, D., Toninello, A., Rugolo, M., Zoccarato, F.: Respiration-linked efflux of Mg^{2+} and other ions from rat liver mitochondria: dependence on Ca^{2+} cycling. In: Mechanisms of proton and calcium pumps. Avron, M., Azzone, G.F., Metcalfe, J., Siliprandi, N. (eds.), pp.263-271. Amsterdam: Elsevier/Excerpta Medica/North-Holland 1978

Siliprandi, D., Toninello, A., Zoccarato, F., Bindoli, A.: Phosphate transport across the mitochondrial membrane: the influence of thiol oxidation and of Mg^{++} on inhibition by mercurials. FEBS Lett. 51, 15-17 (1975a)

Siliprandi, D., Toninello, A., Zoccarato, F., Rugolo, M., Siliprandi, N.: Synergistic action of calcium ions and diamide on mitochondrial swelling. Biochem. Biophys. Res. Commun. 66, 956-961 (1975b)

Siliprandi, D., Toninello, A., Zoccarato, F., Siliprandi, N.: A possible mechanism for respiration-dependent efflux of Mg ions from liver mitochondria. Biochem. Biophys. Res. Commun. 78, 23-27 (1977)

Tietze, F.: Enzymic method for quantitative determination of nanogram amounts of total and oxidized glutathione: application to mammalian blood and other tissues. Anal. Biochem. 27, 502-522 (1969)

Vignais, P.M., Vignais, P.V.: Fucsin, an inhibitor of mitochondrial SH-dependent transport-linked functions. Biochim. Biophys. Acta 325, 357-374 (1972)

Zoccarato, F., Rugolo, M., Siliprandi, D.: Phosphate transport in mitochondrial and submitochondrial particles: the influence of thiol oxidation. J. Bioenerg. and Biomembr. 9, 203-212 (1977)

The Roles of Glutathione, Thioltransferase, and Glutathione Reductase in the Scission of Sulfur-Sulfur Bonds

B. MANNERVIK and K. AXELSSON

Summary

The known pyridine-nucleotide-dependent reductases, lipoamide dehydrogenase, thioredoxin reductase, and glutathione reductase, are highly substrate-specific and do not catalyze reduction of sulfur-sulfur bonds of the wide variety of naturally occurring compounds which contain such bonds. In contrast, cytoplasmic thioltransferase, which utilizes GSH as a reductant, can catalyze a variety of reactions such as reduction of low-molecular-weight disulfides and thiosulfate esters, formation and decomposition of protein-mixed disulfides, and reduction of disulfide bonds in denatured and proteolytically modified proteins. In the reductions NADPH and glutathione reductase are eventually required for the regeneration of GSH from GSSG, which is formed as a product. Thus, the role of GSH in scission of sulfur-sulfur bonds is catalytic.

The thioltransferase has been purified to homogeneity from rat liver cytoplasm and characterized.

Introduction

Sulfur-sulfur bonds in chemical compounds, which occur in vivo, are to a large extent derivable, directly or indirectly, from the amino acid cysteine. Some low molecular weight compounds are simple derivatives of cysteine, such as the disulfide form cystine and mixed disulfides of cysteine and other thiols. Others are derived from glutathione and coenzyme A, which both contain a sulfhydryl group originating from cysteine. In addition to disulfides there is also a group of naturally occurring S-sulfo derivatives of thiols (thiosulfate esters), which contain sulfur-sulfur bonds of the kind found in inorganic thiosulfate. S-sulfoglutathione and S-sulfocysteine are examples of such thiosulfate esters. The occurrence of the above-mentioned low molecular weight compounds containing sulfur-sulfur bonds has been reviewed earlier (1).

A very large class of compounds containing sulfur-sulfur bonds is composed of proteins and polypeptides, in which cysteine residues form disulfide bridges. Many of these disulfide bonds are buried in the interior of the macromolecule, and are thereby inaccessible unless the macromolecule is denatured. The finding that a substantial portion of cytoplasmic glutathione is bound to proteins in the form of mixed disulfides raises questions about the biological function of such more labile disulfides. In Ehrlich ascites tumor cells, 35% of the total GSH was bound to proteins (2) and similar proportions were found in several rat tissues (3,4), as well as in the lens of the eye from several species (5,6). It has also been shown that the level of mixed disulfides of protein and GSH of dormant conidia of the fungus *Neurospora crassa* decreased after germination (7) and that a similar decrease of mixed disulfides of proteins and coenzyme A can be observed in germinating spores of the bacterium *Bacillus megaterium* (8). Furthermore, numerous investigations of isolated proteins have demonstrated that modification of free sulfhydryl groups or reduction of disulfide bonds may modify their biological activities.

In a recent study a regular diurnal variation of the concentrations of GSH, protein sulfhydryl groups, and mixed disulfides of proteins and GSH was established, and the suggestion was put forward that this variation is important for the maintenance of a suitable disulfide/thiol ratio of proteins and thereby to provide protection against both oxidative and reductive stresses (9).

Accordingly, it is evident that biochemical mechanisms must exist for the scission of sulfur-sulfur bonds irrespective of whether such bonds have been formed by adventitious oxidative events or if they are constituents of a regulatory system in the cell. The present communication will review what is known about such mechanisms in liver cytoplasm.

Glutathione Reductase

Three well-characterized enzymes catalyzing reduction of disulfides by NADH or NADPH are known: lipoamide dehydrogenase, thioredoxin reductase, and glutathione reductase (10). Each of these enzymes has a high substrate specificity and cannot be expected to reduce directly sulfur-sulfur bonds in general. Lipoamide dehydrogenase is localized in mitochondria, but it has nevertheless been shown to be capable of reducing some disulfides which can penetrate the mitochondrial inner membrane (11). This reduction probably takes place by means of thiol-disulfide interchange with dihydrolipoyl residues in the multienzyme complexes containing lipoamide dehydrogenase. In the cytoplasm reduced thioredoxin, in a similar manner, can reduce sulfur-sulfur bonds (cf. (12)), but there is no direct information on the importance of this mechanism in relation to the GSH-dependent reduction to be discussed below.

Glutathione reductase is the reductase which has been most definitely associated with the reduction of sulfur-sulfur bonds in general. However, in spite of several claims of lack of specificity, there is no solid evidence for direct reduction of any other naturally occurring disulfides than glutathione disulfide (GSSG) and the mixed disulfide of coenzyme A and GSH. It appears that the lack of significant reactivity of other putative substrates is due to poor binding, because the mixed disulfide of pantetheine and GSH has a high apparent K_m value (13) and higher concentrations of other GSH sulfenyl derivatives give relatively better activities in relation to GSSG than low concentrations (14). The claim that glutathione reductase can reduce mixed disulfides of GSH and hemoglobin or crystallin (15,16) needs further support in view of the finding that direct measurements on the well-defined mixed disulfide of egg-white lysozyme and GSH gave no detectable activity (17). It will appear from the following discussion that the major function of glutathione reductase in reduction of sulfur-sulfur bonds in general is the regeneration of GSH which is used as the reductant. Further information about glutathione reductase can be found in the review by Williams (10); the structure has been solved at 0.3 nm resolution by X-ray diffraction analysis (18) and the catalytic mechanism has been discussed in some detail (19,20).

Thioltransferase

In a study in our laboratory on the reduction of the mixed disulfide of cysteine and GSH an enzyme in rat liver cytoplasm catalyzing the reduction of disulfides by GSH was found (21,22). This enzyme, which probably corresponds to the enzymes catalyzing thiol-disulfide interchange previously described in bovine tissues (23,24), was named thioltransferase in accordance with the nature of the reaction catalyzed (1,25). Previously, this and similar enzymes had been referred to as "transhydrogenases" (23). Some of the earlier investigations on the cytoplasmic thioltransferase, including the reduction of thiosulfate esters and sulfitolysis of GSSG (26), have been reviewed (1). The enzyme has recently been purified to apparent homogeneity and been found to contain 8.6% carbohydrate, have an isoelectric point at pH 9.6, and a molecular weight of about 11,000 (27).

The kinetics of the thioltransferase were investigated with a constant GSH concentration (4 mM) and a variety of substrates containing scissile sulfur-sulfur bonds (Table 1). The reactions were followed by coupling with glutathione reductase, and under the conditions investigated different mixed disulfides of GSH, including the mixed disulfide of lysozyme and GSH (17), and S-sulfoglutathione were the best substrates. Insulin was also among the best substrates and was better by

Table 1. Kinetic constants for thioltransferase-catalyzed and spontaneous reactions between glutathione and disulfides and thiolsulfate esters [a]. [Data from (27)]

Reactant	Apparent parameters of enzyme reaction		Apparent second order rate constant of spontaneous reaction
	K_m (µM)	$10^{-4} \times k_{cat}$ (min^{-1})	k ($mM^{-1}min^{-1}$)
Mixed disulfide of GSH and			
coenzyme A	41	1.20	1.14
L-cysteine	50	0.98	0.22
cysteamine	47	0.73	0.86
egg-white lysozyme	7	0.18	0.44
S-Sulfoglutathione	77	0.85	0.96
S-Sulfocysteine	375	0.31	0.025
L-Cystine	238	0.62	0.049
L-Homocystine	325	0.21	0.012
Cystamine	240	0.59	0.11
O-Benzoylthiamine propyl disulfide	515	0.89	0.025
Trypsin	307	0.36	0.025
Ribonuclease	302	0.19	0.007
Oxytocin	1725	0.33	0.003
Insulin [b]	30	0.58	0.42

[a] GSH was used at 4.05 mM concentration and thioltransferase at 0.394 unit/ml (50 nM) in the standard assay system (30°C) (27).

[b] The buffer of the assay system was changed to 0.14 M Tris-HCl (pH 8.2) instead of 0.14 M phosphate (pH 7.6) used in the standard assay.

a factor > 10 than trypsin, ribonuclease or oxytocin as expressed by the apparent k_{cat}/K_m. With respect to GSH concentration the velocity showed a maximum at about 4 mM GSH when tested with about 0.1 mM S-sulfoglutathione or mixed disulfide of cysteine and GSH (27). The pH-activity profile showed a distinct optimum at about pH 7.5 (27).

All the thioltransferase-catalyzed reactions studied so far can take place in the absence of enzyme, but is has been estimated by extrapolation to the conditions in the cytoplasm that the enzymatic reactions of the mixed disulfide of coenzyme A and GSH (28), as well as other substrates (29), are at least 100 times as rapid as the corresponding spontaneous reactions. In the case of the mixed disulfide of coenzyme A and GSH it has also been calculated that the thioltransferase-catalyzed reaction is more than 20 times more efficient than the direct reduction by glutathione reductase (28), in spite of the fact that this disulfide is the best substrate for glutathione reductase of all mixed disulfides tested. We therefore conclude that direct reduction of sulfur-sulfur bonds catalyzed by glutathione reductase is negligible in vivo for compounds other than GSSG.

We have also considered that the thioltransferase might be involved in the catabolism of disulfide-containing proteins and polypeptides. The nonenzymatic reduction by GSH of some disulfide bonds in proteins has earlier been investigated, and it was found that only few bonds were sterically available for reaction unless the protein had been denatured or subjected to proteolysis (30). Similar experiments with thioltransferase corroborated the published results and demonstrated that the rate of disulfide reduction was enhanced as for the substrates discussed above. The thioltransferase did not appear to increase the number of accessible disulfide bonds, but only to increase the rate of scission of the bonds available for the nonenzymatic reaction. Table 2 shows results from an experiment with bovine serum albumin; other disulfide-containing proteins such as ribonuclease and lysozyme have also been tested. Thus, denaturation of proteins, e.g., by contact with membranes of the cell (31) or by other means, including proteolysis, has to precede the possible action of thioltransferase on protein disulfide groups in the catabolism of these macromolecules.

Finally, in view of the reversible fluctuations of the concentration of mixed disulfides of proteins and GSH (9), we have studied the effect of purified thioltransferase on the following reactions:

1. Protein—SH + GSSG* ⟶ Protein—SSG* + GSH
2. Protein—SSG* + GSH ⟶ Protein—SH + GSSG*

in which the asterisk denotes a radioactively labeled GSH moiety. A crude protein fraction from rat liver, from which endogeneous thioltransferase and GSH had been removed by gel filtration on Sephadex G-75 (cf. (29)), was used. For reaction 1 the protein was pretreated with dithioerythritol to reduce possible mixed disulfides of the proteins and, after removal of the reagent, assayed with radioactive GSSG and purified thioltransferase. For reaction 2 labeled mixed disulfide was prepared from reduced protein and labeled GSSG (according to reaction 1). The labeled protein was then assayed with unlabeled GSH and thioltransferase. In both cases it was demonstrated that the presence of thioltransferase increased the reaction rate significantly.

In conclusion it can be stated that the enzyme system consisting of GSH and thioltransferase coupled to glutathione reductase is capable of catalyzing a large variety of reactions involving scission of sulfur-sulfur bonds. The majority of the compounds containing scissile sulfur-

Table 2. Effect of proteolysis on reduction of disulfide groups in bovine serum albumin [a]

Protease	No. of SS bonds reduced	Velocity (µM/min)	
		Spontaneous	Thioltransferase-catalyzed
Control	0.68	0.9	1.3
Trypsin	5.9	1.5	12.1
Chymotrypsin	5.8	1.0	10.2
Pronase	6.2	1.4	10.0
Trypsin + chymotrypsin + pronase	10.6	2.2	14.7

[a] Bovine serum albumin (0.14 mM) was preincubated with the proteases (0.4 µM each) for 20 h before assay with thioltransferase and 0.5 mM GSH. The samples were diluted 5 times in the assay system.

sulfur bonds are related to cysteine, but it has not so far been demonstrated that the enzyme system has specificity for cysteine derivatives. The reactions include : (a) reduction of low-molecular-weight compounds (e.g., cystine), which may be a prerequisite for the utilization of a moiety in biochemical reactions (e.g., cysteine in protein biosynthesis); (b) formation and decomposition of protein-mixed disulfides, which may buffer changes in the redox status of the cell or serve specific regulatory functions for individual proteins; and (c) reduction of disulfide bonds in the catabolism of proteins and polypeptides. In all these reactions thioltransferase has a crucial role, because glutathione reductase is highly substrate-specific and the spontaneous reactions are small in relation to the thioltransferase-catalyzed reactions.

Acknowledgment. The work described in this paper has been supported by grants (to B.M.) from the Swedish National Science Research Council.

References

1. Mannervik, B., Eriksson, S.A.: In: Glutathione. Flohé, L., Benöhr, H.C., Sies, H., Waller, H.D., Wendel, A. (eds.), pp.120-131. Stuttgart: Georg Thieme 1974
2. Modig, H.: Biochem. Pharmacol. 17, 177-186 (1968)
3. Harrap, K.R., Jackson, R.C., Riches, P.G., Smith, C.A., Hill, B.T.: Biochim. Biophys. Acta 310, 104-110 (1973)
4. Harisch, G., Schole, J.: Z. Naturforsch. 29c, 261-266 (1974)
5. Herrmann, H., Moses, S.G.: J. Biol. Chem. 158, 33-45 (1945)
6. Harding, J.J.: Biochem. J. 117, 957-960 (1970)
7. Fahey, R.C., Brody, S., Mikolajczyk, S.D.: J. Bacteriol. 121, 144-151 (1975)
8. Setlow, B., Setlow, P.: Biochem. Biophys. Res. Commun. 75, 45-50 (1977)
9. Isaacs, J., Binkley, F.: Biochim. Biophys. Acta 497, 192-204 (1977)
10. Williams, C.H., Jr.: Enzymes, 3rd ed. 13, 89-173 (1976)
11. Eldjarn, L., Bremer, J.: Acta Chem. Scand. 17, Suppl.1, 59-66 (1963)
12. Holmgren, A.: J. Biol. Chem. 252, 4600-4606 (1977)
13. Mannervik, B., Nise, G.: Arch. Biochem. Biophys. 134, 90-94 (1969)
14. Carlberg, I., Mannervik, B.: J. Biol. Chem. 250, 5475-5480 (1975)

15. Srivastava, S.K., Beutler, E.: Biochem. J. 119, 353-357 (1970)
16. Srivastava, S.K., Beutler, E.: Exp. Eye Res. 17, 33-42 (1973)
17. Axelsson, K., Mannervik, B.: FEBS Lett. 53, 40-43 (1975)
18. Schulz, G.E., Schirmer, R.H., Sachsenheimer, W., Pai, E.F.: Nature (London) 273, 120-124 (1978)
19. Williams, C.H., Jr., Thorpe, C., Arscott, L.D.: In: Mechanisms of oxidizing enzymes. Singer, T.P., Ondarza, R.N. (eds.). New York: Elsevier 1978 (in press)
20. Mannervik, B., Boggaram, V., Carlberg, I., Larson, K.: In: Flavins and Flavoproteins. Yagi, K., Yamano, T. (eds.). Tokyo: Japan Scientific Societies Press 1978 (in press)
21. Eriksson, B., Eriksson, S.A.: Acta Chem. Scand. 21, 1304-1312 (1967)
22. Eriksson, S.A., Mannervik, B.: FEBS Lett. 7, 26-28 (1970)
23. Racker, E.: J. Biol. Chem. 217, 867-874 (1955)
24. Chang, S.H., Wilken, D.R.: J. Biol. Chem. 241, 4251-4260 (1966)
25. Askelöf, P., Axelsson, K., Eriksson, S., Mannervik, B.: FEBS Lett. 38, 263-267 (1974)
26. Mannervik, B., Persson, G., Eriksson, S.: Arch. Biochem. Biophys. 163, 283-289 (1974)
27. Axelsson, K., Eriksson, S., Mannervik, B.: Biochemistry 17, 2978-2984 (1978)
28. Eriksson, S., Guthenberg, C., Mannervik, B.: FEBS Lett. 39, 296-300 (1974)
29. Mannervik, B., Axelsson, K.: Biochem. J. 149, 785-788 (1975)
30. Davidson, B.E., Hird, F.J.R.: Biochem. J. 104, 473-479 (1967)
31. Ballard, F.J.: Essays in Biochemistry 13, 1-37 (1977)

Chapter IV
Pharmacological and Toxicological Aspects

The Glutathione Transferases in Detoxification

W. B. JAKOBY

Summary

The glutathione transferases (EC 2.5.1.18) appear to perform several detoxification functions. This group of enzymes is analogous to serum albumin in the broad spectrum of compounds that serve as ligands, but differs significantly in that the transferases have an additional and specific site for GSH. They appear to catalyze all reactions that would be expected of GSH acting as a nucleophile, providing that the electrophilic substrate is bound to the enzyme. Examples of the reactions catalyzed include conjugation to compounds with a reactive electrophilic carbon to form thioethers; reactions with other electrophilic atoms including the nitrogen of organic nitrates and the sulfur of organic thiocyanates; isomerization; disulfide interchange; glutathione peroxidase activity with organic peroxides; and the formation of a thiolester from p-nitrophenylacetate.

The enzymes have been shown to act as storage proteins and appear to be responsible for the binding of bilirubin and its conjugates, among many other compounds, within the hepatocyte. A more speculative scavenger function has also been suggested for them.

These enzymes are widely distributed in higher animals and constitute 10% and 3%, respectively, of the extractable protein of rat and human liver.

A. Introduction

It is encouraging to observe that glutathione, a compound in search of a physiological function for several decades (and during several symposia), is now able to call upon this large number of investigators, each able to offer a separate aspect in answer to the question of physiological role. Our own offering is an enzyme with multiple functions in detoxification (1).

After noting that the concentration of GSH in liver is normally of the order of 5 mM, I will confine myself to our own work with the glutathione transferases (EC 2.5.1.18) and mention that this group of enzymes constitutes 10% and 3%, respectively, of the total extractable protein from rat (2) and human (3) liver. Having established that there is a large amount of both substrate and enzyme, we ask what the enzyme does. Here we also deal in large numbers because it does many things; the substrates for the protein's catalytic role alone are legion[1]. Although this paper is concerned primarily with the variety of catalytic effects of the glutathione transferases, these proteins justify their high con-

[1] My dictionary tells me that a (Roman) legion was comprised of approximately 3000 to 6000 foot soldiers and cavalry. That seems to account very well for the known number of substrates.

centration in tissue by their additional roles which will be mentioned subsequently, though briefly.

Fortunately, the background of their discovery and the extensive early work with the enzymes has been reviewed (4). In the present context, glutathione transferase refers to membership in a group of enzymes with overlapping specificity that have been purified to homogeneity from both rat and human liver. The details of our purification methods for the five enzymes from rat and the five from man have been presented and the characterization of the individual proteins has been discussed (5). The rat transferases are designated by roman letters in the reverse order of their elution from CM-cellulose and the human transferases are designated by Greek letters in increasing order of their isoelectric points. We know that the time of the glutathione transferases has arrived since there is now a controversy about their kinetics (5-7), a certain sign of significance.

B. Catalytic Role

The range of glutathione transferase activity is encompassed by the statement that the enzyme will catalyze any of the possible nucleophilic reactions of GSH acting as a thiolate anion (8,9). This hypothesis is subject only to the obvious qualification that the enzyme must be able to bind the potential substrate, a premise that will be examined here.

I. Attack at an Electrophilic Carbon

There are hundreds of compounds bearing a sufficiently electrophilic carbon that can react with GSH to form the corresponding thioether of which Reaction 1 is an example. The leaving group may be a chloride

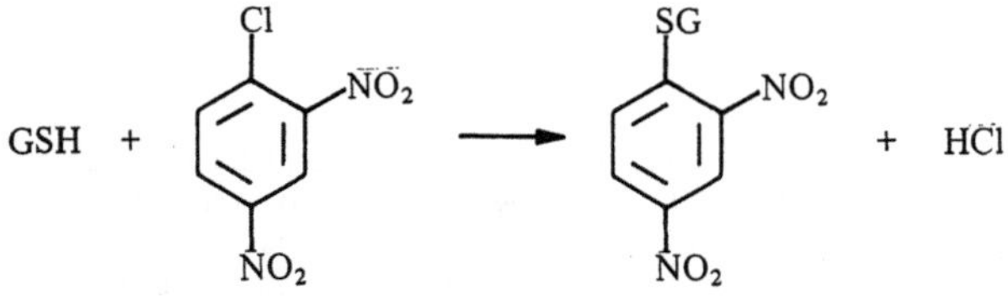

(1)

as shown, but could be nitrite (as in p-nitropyridine N-oxide), bromide (as in bromosulfophthalein), or iodide (as in iodomethane). Similarly, the electrophilic carbon could be present in an arene oxide (as in benzo[a]pyrene 4,5-oxide) or an α,β-unsaturated ketone (as in trans-4-phenyl-3-buten-2-one). A partial list of such substrates, and the specific activity of the several glutathione transferases for them, have been prepared (5,10).

This type of reaction represents the first step in the formation of the generic group of excretion products known as the mercapturic acids. The conjugate with GSH may then undergo transpeptidation resulting in the removal of the γ-glutamyl group (11) and, subsequently, hydrolysis of glycine. The resultant product can be N-acetylated to form the N-acetylcysteine thioether of the original electrophilic compound which is designated as a mercapturic acid.

From the small number of substrates already alluded to, and from the large number known from work with the enzymes (5,10), it will be evident that there is little here in terms of specificity except that for an electrophilic carbon and the requirement for a hydrophobic aspect to the substrate molecule. In quantitative terms, the effect of electrophilicity on reaction rate has been evaluated by use of Hammett plots for a group of 1-chloro-2-nitrobenzenes substituted in position 4. The data were entirely in accord with the conclusion that greater velocity in the enzyme-catalyzed reaction occurred for those compounds bearing para substituents of increasing electron-withdrawing capacity as described by their σ^- constants (8). The major specificity of the enzyme appears to reside in the requirement for GSH itself as the nucleophile (10); cysteine or other thiols do not replace GSH.

II. Attack at Other Electrophilic Atoms

In view of the lack of specificity, it appeared reasonable to consider that even the electrophilic carbon need not be an absolute requirement for catalysis and this proved to be the case. As illustrated in Figure 1, the nitrogen of organic nitrate esters and the sulfur of thiocyanates are examples of effective targets for nucleophilic attack by GSH (8). When other possible type-reactions were examined, the transferases were found to catalyze isomerization of a double bond system (12) as shown by the formation of α,β-unsaturated Δ^4-3-ketosteroids from Δ^5-3-ketosteroids (Fig.1). The transferases were also capable of allowing thioester formation from GSH and p-nitrophenylacetate (Fig.1) (9). Recently, glutathione peroxidase activity in an enzyme free of selenium was identified as an additional function of the transferases (13); homogeneous preparations of transferase B have been shown by T.C. Stadtman to be particularly high in this activity which is ef-

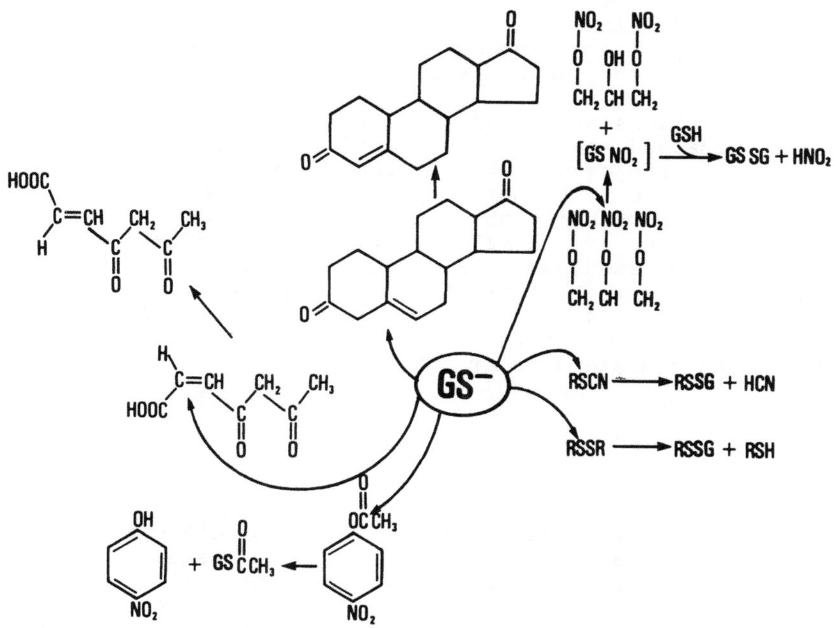

Fig. 1. Scheme of some of the reactions in which glutathione thiolate is active. Conjugation reactions are not shown

Table 1. Specific activity for substrates of the glutathione transferase

Substrate	GSH-S-tranferases from rat					Human (range)
	AA	A	B	C	E	
	μmol/min/mg					
1,2-Dichloro-4-nitrobenzene	0.008	4.3	0.003	2.0	<0.001	0.025-0.03
1-Chloro-2,4-dinitrobenzene	14	62	11	10	0.01	16-37
Benzo[a]pyrene-4,5-oxide	0.087	0.011	<0.098	0.069	0.004	0.025-0.030
Iodomethane	1.4	<0.02	0.59	<0.01	8.9	13
Ethacrynic acid	0.3	<0.01	0.26	0.11	<0.01	0.017-0.044
Bromosulfophthalein	0.004	0.53	0.006	0.18	<0.001	0.001-0.010
Trinitroglycerol	0.14	0.06	0.09	0.37		0.05 -0.32
Ethylthiocyanate	0.02	0.08	0.02	0.57		0.13 -0.28
Maleylacetone	0.07	0.06	0.08	0.06		0.15
p-Nitrophenyl acetate	0.27	0.56	0.33	0.36		15.0
Cystine dimethyl ester	0.14	0.06	0.01			0.21

fective only with organic peroxides (25). The specific activity of several of the glutathione transferases with a selected few substrates is presented in Table 1.

III. Contradictions and Resolution

In accord with our hypothesis, the known spontaneous and enzyme catalyzed (14) mercaptan-mediated isomerization of maleic acid to form fumaric acid should be a function of the transferases; it is not! Similarly, the rate of the disulfide interchange reactions should be affected by the enzyme, but there is no change in rate of the reaction of cystine and GSH to form the mixed disulfide.

These apparent contradictions were placed in perspective when it was considered that an uncharged analog, maleic acid dimethyl ester, does serve as a substrate for conjugation by the transferases. Maleylacetoacetate (Fig.1) and maleylacetone, both representing an intermediate position in that they both possess ionized groups, were found to be substrates for isomerization (9). Similarly, when the anionic functional groups of cysteine were blocked, as in N,N-diacetyl-L-cysteine, the resultant derivative became a substrate for disulfide interchange with GSH; reaction of GSH with uncharged aromatic disulfides also occurred readily (9). It has already been noted above that the glutathione peroxidase activity of the transferases is not effective with hydrogen peroxide (13).

IV. Mechanism

Our present view of the catalytic mechanism envisages the glutathione transferases as binding proteins with a second site for GSH close by. Since the role of GSH is clearly that of a nucleophile, catalysis seems to depend on only two major attributes of substrate: that it binds at an active site and that it bears a sufficiently electrophilic atom to allow reaction with GSH. Although the number and variety of compounds that fulfill this requirement are enormous, it is necessary to evaluate the efficiency of an arrangement with such apparently simple requirements. For one of the best substrates, identified in Reaction 1, the rate enhancement[2] due to the enzyme is approximately 500 M (8). This small value is consistent with the conclusion that catalytic action simply reflects the proximity of GSH to that of the second substrate. Such a primitive mechanism has its mechanical counterpart in nothing more complex than a wedge and, indeed, the transferases are not very efficient catalysts.

C. Other Detoxification Functions

Aside from the large range of substrates that must perforce bind to the transferases prior to catalysis, there is another, probably larger group of compounds that also act as ligands but not as substrates for the enzyme. In fact the original discovery of glutathione transferase B took place as the result of a search by several investigators for a binding protein effective with bilirubin (16), with a metabolite of cortisol (17), and with an azo-dye carcinogen (18). After independent isolation of proteins capable of each of these binding functions, it was possible to show that a single protein, named ligandin, was responsible for each of the enumerated activities (19). Only later were we able to ascertain that ligandin was identical to glutathione transferase B (20) and this demonstration led to the question of whether such avid binding was an intrinsic property of the transferases. The data are entirely consistent with the view that each of the glutathione transferases is a binding protein for compounds bearing a sufficiently large hydrophobic domain (21). Included, of course, are all the substrates varying from iodomethane to the large epoxides of the polycyclic aromatic hydrocarbons. In addition are such nonsubstrate as heme, bilirubin and its glucuronides, and many of the dyes used in the assessment of liver function (21); all are inhibitors of the catalytic activity (21).

I. Storage

This strictly binding function appears to be of major importance in the intracellular storage of the large number of xenobiotics including drugs that pass through the liver, as well as of bilirubin, the toxic metabolite of heme that is produced at a rate of 250 mg per day in man. Our recent work is in accord with the transferases as intrahepatic storage proteins (22,23). The need for a storage protein will be evident if it is recognized that bilirubin is essentially insoluble under physiological conditions, and would simply form a sludge if not bound to a soluble matrix; in the circulation, this role is performed by albumin and in the hepatocyte by the transferases. Thus, a large number of compounds can be quickly removed from the circulation and stored in the liver in a bound form while awaiting metabolic conversion to less toxic compounds.

II. A Scavenger Role

There is increasingly convincing data that the highly electrophilic products of oxidative processes, e.g., the reactive epoxides formed

[2]This is taken as the effective concentration of the aromatic substrate at the reactive site of the E-GSH complex, necessary to account for the observed rate in the presence of enzyme (15). Rate enhancement = k_{cat}/k_2.

by the P-450 mixed function oxidases may play a major role in carcinogenesis (24). Some of these compounds are detoxified by the transferases in that they are bound by these enzymes and then react with GSH. But once bound to a transferase, sufficiently reactive electrophiles also have the option of reacting with nucleophilic groups on the protein itself. For example, in the absence of GSH, the substrate 1-chloro-2,4-dinitrobenzene rapidly forms a covalent bond with the transferases leading to 0.3 mol of reagent bound per mol of enzyme after incubation for 5 min.

We have ascribed a scavenger function to the transferases because of this penchant for suicide. It should not be surprising that some of the substrates, those that are particularly vigorous alkylating agents, interact with protein leading to inactivation; many proteins behave similarly (24). However, by reason of their avid binding, the transferases are able to position the reactive compounds on the protein and thereby provide increased opportunity for reaction. It is of interest that one of the three assays leading to the isolation of ligandin was based on the yellow color that was due to *covalent* linkage of an azo-dye carcinogen (butter yellow) to the protein (18).

D. Comments

Although our prime interest here has been to present a mechanism for the catalytic functions of the glutathione transferases in catalyzing nucleophilic reactions of GSH, it will be clear that these proteins perform a number of roles, not limited to liver (5), in protecting the animal from the enormous number of noxious compounds that are inhaled, ingested, and produced metabolically. Since the concentration of transferases is so high, even the admittedly speculative sacrificial role that has been outlined can be accommodated. Although these proteins appear to be primitive enzymes and relatively unspecific binding proteins, it may be precisely this simplistic design that allows such protean serviceability in their role in detoxification.

References

1. Jakoby, W.B., Keen, J.H.: Trends Biochem. Sci. 2, 229-231 (1977)
2. Jakoby, W.B., Ketley, J.N., Habig, W.H.: In: Glutathione: metabolism and function. Arias, I.M., Jakoby, W.B. (eds.), pp.213-223. New York: Raven Press 1976
3. Kamisaka, K., Habig, W.H., Ketley, J.N., Arias, I.M., Jakoby, W.B.: Eur. J. Biochem. 60, 153-161 (1975)
4. Boyland, E., Chasseaud, L.F.: Adv. Enzymol. 32, 172-219 (1969)
5. Jakoby, W.B.: Adv. Enzymol. 46, 383-414 (1978)
6. Mannervik, B., Askelof, P.: FEBS Lett. 56, 218 (1975)
7. Pabst, M.J., Habig, W.H., Jakoby, W.B.: J. Biol. Chem. 249, 7140-7150 (1974)
8. Keen, J.H., Habig, W.H., Jakoby, W.B.: J. Biol. Chem. 251, 6283-6288 (1976)
9. Keen, J.H., Jakoby, W.B.: J. Biol. Chem. 253, 5654-5657 (1978)
10. Jakoby, W.B., Habig, W.H., Keen, J.H., Ketley, J.N., Pabst, M.J.: In: Glutathione: metabolism and function. Arias, I.M., Jakoby, W.B. (eds.), pp.189-211. New York: Raven Press 1976
11. Tate, S.S., Meister, A.: J. Biol. Chem. 249, 7593-7602 (1974)
12. Benson, A.M., Talalay, P., Keen, J.H., Jakoby, W.B.: Proc. Nat. Acad. Sci. U.S.A. 74, 158-162 (1977)
13. Prohaska, J.R., Ganther, H.E.: Biochem. Biophys. Res. Commun. 76, 437-445 (1977)
14. Scher, W., Jakoby, W.B.: J. Biol. Chem. 244, 1878-1882 (1969)

15. Jencks, W.P.: Catalysis in chemistry and enzymology, pp.4-5. New York: McGraw-Hill 1969
16. Reyes, A., Levi, A.J., Arias, I.M.: J. Clin. Invest. 50, 2242-2252 (1971)
17. Morey, K.S., Litwack, G.: Biochemistry 8, 4813-4821 (1969)
18. Ketterer, B., Ross-Mansell, P., Whitehead, J.K.: Biochem. J. 103, 316-324 (1967)
19. Litwack, G., Ketterer, B., Arias, I.M.: Nature (London) 234, 466-467 (1971)
20. Habig, W.H., Pabst, M.J., Fleischner, G., Gatmaitan, Z., Arias, I.M., Jakoby, W.B.: Proc. Nat. Acad. Sci. U.S.A. 71, 3879-3882 (1974)
21. Ketley, J.N., Habig, W.H., Jakoby, W.B.: J. Biol. Chem. 250, 8670-8673 (1975)
22. Wolkoff, A.W., Ketley, J.N., Waggoner, J.G., Berk, P.D., Jakoby, W.B.: J. Clin. Invest. 61, 142-149 (1978)
23. Wolkoff, A.W., Weisiger, R.A., Jakoby, W.B.: In: Progress in liver diseases. Popper, H., Schassner, F. (eds.), Vol.6. In press (1979)
24. Miller, E.C., Miller, J.A.: In: Molecular biology of cancer. Busch, H. (ed.), pp.377-402. New York: Academic Press 1974
25. Stadtman, T.C.: Nutrition Reviews 35, 161 (1977)

Functions of Glutathione in Drug Metabolism

S. ORRENIUS and D. P. JONES

Summary

The liver cytochrome P-450 monooxygenase system can be responsible for generation of a variety of reactive metabolites, including hydrogen peroxide, aldehydes, and epoxides which, unless inactivated by further metabolism, may produce cytotoxicity. In this study we have examined the role of glutathione (GSH) in detoxification of such reactive species in isolated hepytocytes by investigating (1) the role of glutathione peroxidase in H_2O_2 decomposition, (2) the role of formaldehyde dehydrogenase in HCHO oxidation, and (3) the role of GSH in bromobenzene metabolism.

The level of catalase Compound I in isolated hepatocytes was not increased by addition of ethylmorphine or aminopyrine; however, a rapid oxidation of GSH occurred. Depletion of GSH by pretreatment of the rats with diethylmaleate resulted in no increase in the basal rate of H_2O_2 production by isolated hepatocytes, but addition of a drug substrate to these cells approximately doubled the estimated rate of peroxide production. Therefore, it appears that without added substrate only a small amount of H_2O_2 may be produced by the cytochrome P-450 monooxygenase. Addition of a drug substrate, however, stimulates the generation of H_2O_2, which is subsequently detoxified primarily by a GSH-dependent process, presumably glutathione peroxidase.

N-demethylation of drugs by the hepatic cytochrome P-450 system results in rapid production of HCHO, which is subsequently oxidized to HCOOH, CO_2 and H_2O. The accumulation of a small concentration of formic acid during N-demethylation was detected by its titration of catalase Compound I and subsequent binding as an anionic ligand. Depletion of GSH by pretreatment of rats with diethylmaleate slowed these processes, which indicated a role for GSH-dependent formaldehyde dehydrogenase in HCHO oxidation.

Reactive epoxides, generated from bromobenzene by the cytochrome P-450 monooxygenase, appear to be responsible for the hepatotoxicity of this drug. In isolated hepatocytes, bromobenzene toxicity is seen only following depletion of GSH either by pretreatment of the rats with diethylmaleate or when incubation with bromobenzene is performed in a noncomplete medium. The presence of methionine or cysteine in the medium supports resynthesis of GSH and prevents cell death. Addition of GSH to microsomal incubations with bromobenzene prevents binding of bromobenzene metabolites to protein. Thus, GSH functions as a nucleophile in combinations with electrophilic bromobenzene metabolites to prevent their cytotoxicity in hepatocytes.

This study demonstrates that in isolated hepatocytes GSH plays an important role in detoxication of reactive species produced by the cytochrome P-450 system in three characteristic manners: (1) as a reducing agent, (2) as a cofactor, and (3) as a nucleophile in conjugation reactions.

A. Introduction

The sulfhydryl group of glutathione has received considerable attention in the search for functions of this tripeptide. This has occurred for two reasons: first, the oxidation-reduction potential at neutral pH is about -230 mV, a value which is of physiological interest, since it suggests that glutathione can function in electron transfer reactions with a variety of cellular constituents, and second, the reduced form (GSH) reacts as a nucleophile with a number of electrophilic substances. This work has led to the discovery of numerous glutathione-dependent oxido-reductases (3) and GSH transferases (7) which utilize these properties. Many of the enzymes appear to be important in detoxication of foreign substances.

Cytochrome P-450, an enzyme system that has been recognized since its discovery as playing an important role in the metabolism and detoxication of a variety of lipophilic foreign compounds, has more recently been found rather efficient also in generation of toxic products. Thus, it catalyzes the formation of reactive oxygen species (O_2^- and H_2O_2), aldehydes (HCHO and CH_3CHO), epoxides, and other reactive electrophiles.

During the last few years we have used suspensions of freshly isolated hepatocytes for study of drug metabolism and toxicity, and have more recently utilized this model also to examine the role of glutathione in detoxication of reactive products generated by the cytochrome P-450 monooxygenase (9,21-23). We will here describe three different ways in which GSH may be used in detoxication reactions, i.e., GSH can be used as a reductant, GSH can be used as a cofactor, or GSH can be used as a nucleophile in conjugation of electrophiles.

In the first of these reactions, catalyzed by glutathione peroxidase, the oxidation-reduction properties of glutathione are used to affect the reduction of peroxides, especially hydrogen peroxide, which may be generated during drug metabolism presumably due to release of O_2^- as a result of dissociation of the cytochrome P-450-oxy complex (16). In this reaction GSH is oxidized to GSSG with the concomitant reduction of the peroxide. The second type of reaction concerns the GSH-dependent formaldehyde dehydrogenase, a reaction system in which GSH is required as a cofactor for oxidation of formaldehyde to formic acid (20). In this reaction sequence, GSH is neither oxidized nor consumed. In the third type of reaction, which may be catalyzed by GSH transferases or occur nonenzymatically, GSH reacts with electrophilic products to form conjugates which are typically less toxic than the original compound and also more readily eliminated from the body, usually as mercapturic acid derivatives. In the latter type of reaction, GSH is lost from the cell, and if the rate of generation of electrophiles is more rapid than resynthesis of GSH, it may result in binding to other cellular components and subsequent cellular damage.

B. The Role of Glutathione Peroxidase in Decomposition of H_2O_2 Produced during Drug Metabolism in Isolated Hepatocytes

Gillette and coworkers demonstrated in 1957 that H_2O_2 is produced upon aerobic incubation of microsomes in the presence of NADPH (4). This production is now recognized to be due to a combination of autooxidation of the flavoprotein NADPH-cytochrome P-450 reductase and dissociation of the cytochrome P-450-oxy complex (1,16,24). The direct involvement of cytochrome P-450 (oxidase function) in H_2O_2 (or O_2^-) production by liver microsomes is well established and has been shown to be dependent on both substrate (12) and induction scheme (6). However, the significance of H_2O_2 production in the endoplasmic reticulum in intact tissues has been questioned (13,17) primarily because the rate of H_2O_2 production measured by the catalase Compound I method on perfused liver is much lower than that expected on the basis of microsomal H_2O_2 production, and because this rate is not increased by pretreatment with phenobarbital or in the presence of aminopyrine. On the other hand, substrates of cytochrome P-450 result in release of GSSG from perfused liver (13,17) just as is observed when substrates of glutathione peroxidase are introduced (17). Consequently, the possibility remained that drug-induced H_2O_2 production does occur also in intact tissue, but that any H_2O_2 produced is reduced by glutathione peroxidase and not detected by the catalase method. We examined this hypothesis by using isolated hepatocytes and measuring both H_2O_2 production and the glutathione pool.

To measure H_2O_2 production in intact cells we used the catalase Compound I titration developed by Sies and Chance (14,18,19). With this method the rate of peroxide production is proportional to the concentration of methanol ($a_{1/2}$) required for a 50% decrease in the steady-state concentration of the H_2O_2-catalase complex termed Compound I. A typical titration, obtained with cells from a phenobarbital pretreated rat, is shown in Figure 1. The endogenous concentration of Compound I represented about 40%-50% of total catalase as measured by KCN titration (13).

Rates of H_2O_2 production without added drug substrates, and with glycolate or octanoate (9) corresponded to values obtained in perfused liver (13). Further, as previously reported for perfused liver (13), addition of drugs such as aminopyrine or ethylmorphine had no effect on the mea-

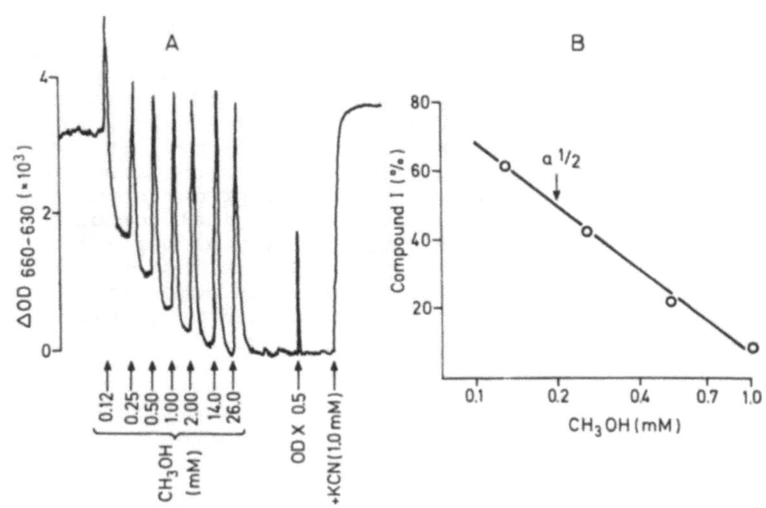

Fig. 1A and B. Catalase Compound I titration with CH_3OH in hepatocytes from phenobarbital pretreated rats. Conditions: 1.7×10^6 cells/ml in Krebs-Henseleit buffer containing 25 mM Hepes, pH 7.4; light path 3.9 cm. (A) Absorbance change. (B) Percent Compound I remaining during titration. Data from (9)

Table 1. Rates of H_2O_2 production and GSH oxidation in isolated hepatocytes [a]

Pretreatment	Additions	H_2O_2 production [b] (nmol/10^6 cells/min)	GSSG excretion [c] (nmol/10^6 cells/min)
None	None	1.52	-
Phenobarbital [d]	None	1.32	0.12
	Ethylmorphine (1 mM)	1.32	0.87
	Aminopyrine (5 mM)	1.29	0.48
	Perfluoro-n-hexane (0.83 mM)	1.41	0.32
Phenobarbital + diethylmaleate [d]	None	1.45	-
	Ethylmorphine (1 mM)	3.71	-
	Aminopyrine (1.1 mM)	3.22	-
	Perfluoro-n-hexane (0.83 mM)	2.91	-

[a] Data for no additions and ethylmorphine are from (9).
[b] Measured by catalase Compound I titration, see Figure 1.
[c] Measured by fluorometric assay as GSSG formed.
[d] Pretreatment as described in (9).

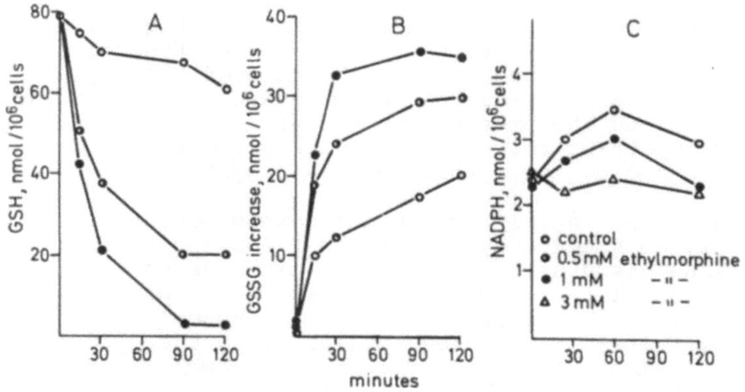

Fig. 2A-C. GSH oxidation and NADPH concentration in hepatocytes in presence of ethylmorphine. (A) GSH concentration in cell pellet. (B) GSSG increase in supernatant, expressed relative to initial cell concentration. (C) NADPH concentration. Data of (A) and (B) from (9)

sured rate of H_2O_2 production, and cells from phenobarbital pretreated rats had a lower rate of H_2O_2 production than those from controls (Table 1).

Addition of these same drugs to hepatocytes from phenobarbital pretreated rats resulted in a rapid decrease in GSH content (Table 1, Fig.2A). This decrease was related to a stoichiometric increase in GSSG (Fig.2B). The increase in GSSG was totally extracellular; no increase in intracellular GSSG was detected. GSH oxidation did not appear to be due simply to a depletion of NADPH, which is required for reduction of GSSG to GSH, since NADPH concentration in hepatocytes was moderately decreased (Fig.2C). It thus appeared that GSH may be oxidized by the glutathione peroxidase reaction coupled to reduction of H_2O_2.

To examine this possiblity we used diethylmaleate (DEM) treatment of rats to deplete the hepatocytes of glutathione. The hypothesis was that if glutathione peroxidase played a major role in decomposition of drug-induced H_2O_2 production, then by depleting the cells of GSH, the increased H_2O_2 production should be detectable by the catalase Compound I system. Depletion of GSH without added substrate had no effect on $a_{1/2}$ (Table 1) suggesting that endogenous H_2O_2 production in the endoplasmic reticulum is small compared to other H_2O_2 sources in the cells, in agreement with conclusions of Oshino and Chance (13). However, when we now added ethylmorphine, aminopyrine, or perfluoro-n-hexane, we found that the $a_{1/2}$ was increased (Table 1). This suggests that drug-induced H_2O_2 production in the endoplasmic reticulum does occur, but that this does not reach catalase which is located in the peroxisomes. Rather, H_2O_2 produced upon addition of drug substrate, appears to be normally decomposed by a GSH-dependent process, presumably glutathione peroxidase.

A comparison of the rate of GSH oxidation in cells from phenobarbital-treated rats in the presence of ethylmorphine (1 mM) and the rate of increased H_2O_2 production in hepatocytes from rats treated with both phenobarbital and DEM is interesting from two points. First, the rate of GSH oxidation is about half the rate of the stimulation of H_2O_2 production, which may reflect that under normal cellular conditions, when GSH is oxidized as a consequence of peroxide reduction, the GSSG produced is not completely reduced by glutathione reductase, but rather much of it is released from the cell. Second, the rate of GSH oxidation and H_2O_2 production induced by certain drugs is very high, i.e., over 2 nmol/10^6 cells/min, which is probably about 25% of the rate of ethylmorphine demethylation and comparable to the rate of many cytochrome P-450-dependent oxidations. Thus, as a consequence of its oxidation-reduction potential and the presence of glutathione peroxidase, GSH appears to have an important role in isolated hepatocytes in decomposition of H_2O_2 produced by the cytochrome P-450 system.

These observations add to the rapidly accumulating knowledge concerning generation of active oxygen species by cytochrome P-450 and their decomposition by superoxide dismutases, glutathione peroxidases and catalase. There is little doubt that H_2O_2 is produced following drug addition to intact cells, and based upon studies with subcellular fractions it appears likely that it is produced primarily by the cytochrome P-450 system. The mechanism of this production has not been finally established; however, the involvement of superoxide anions appears more than likely. In addition, the recent studies using selenium-deficient animals strongly suggest that more than one form of glutathione peroxidase is present in liver (10) and that in addition to drug-induced H_2O_2 production, drugs may also stimulate formation of other peroxides which can act as substrates for glutathione peroxidases.

C. The Role of GSH-Dependent Formaldehyde Dehydrogenase in Decomposition of Formaldehyde Produced by N-Demethylation of Drugs in Isolated Hepatocytes

The GSH-dependent oxidation of formaldehyde to formic acid appears to involve two enzyme-catalyzed reactions for which GSH is required as a cofactor (20). Formaldehyde reacts nonenzymatically with GSH to form S-hydroxymethylglutathione, which is the substrate for formaldehyde dehydrogenase, which catalyzes the transfer of two electrons to NAD^+ and results in formation of formylglutathione. A second enzyme, formylglutathione hydrolase, catalyzes hydrolysis to yield formic acid and

GSH. Although, these enzymes have been purified and characterized (5, 25,26), their function in intact tissues has not been established.

When we began to study the drug-induced H_2O_2 production with the catalase Compound I system, we found that the amount of Compound I which could be titrated with methanol decreased during the incubation of hepatocytes with ethylmorphine or aminopyrine (9). At a constant length of time, the amount of Compound I detected decreased as the concentration of ethylmorphine was increased (Table 2). This decrease appeared to be related in some way to a loss of catalase since the amount of catalase detectable at the end of the experiment by addition of KCN was also decreased. The decrease seemed to be due to combination of catalase with formic acid. Such binding is known to occur readily and the affinity of catalase for formic acid is quite high (2). In addition, formic acid can act as an electron donor to catalase Compound I, so that titration of Compound I may not be directly correlated on a time sequence to loss of KCN detectable catalase.

Since formic acid oxidation is decreased in hepatocytes when methionine is limiting (27) it seemed possible that the interaction of formic acid with catalase was observed only because the cells were depleted of methionine. Consequently, we preincubated hepatocytes with methionine prior to addition of ethylmorphine (Fig.3). This had no effect on the decrease in Compound I, which suggests that the formic acid interaction with catalase occurs even though quantitatively it may represent only a small fraction of the total formic acid oxidation.

When we used cells from rats which had been pretreated with diethylmaleate, we found that the loss of catalase Compound I, and also of KCN detectable catalase, was diminished. Thus, the availability of formic acid for the catalase-dependent reactions was GSH-dependent. This is exactly what one would predict if the GSH-dependent formal-

Table 2. Catalase Compound I remaining after 6 min incubation of hepatocytes with ethylmorphine [a]

Ethylmorphine (mM)	Control	Diethylmaleate pretreated
0.1	60	97
0.2	58	87
1.0	42	68
2.0	32	55
8.0	16	39

[a] Expressed as percentage relative to titration without added ethylmorphine.

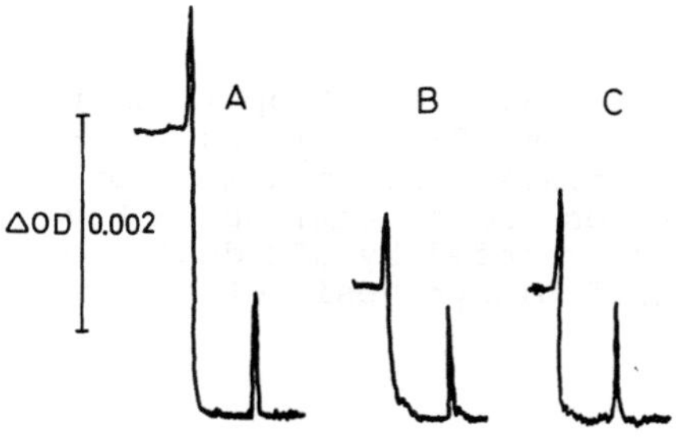

Fig. 3A-C. Effect of methionine on Compound I decrease in the presence of ethylmorphine. (A) Absorbance change (660 minus 630 nm) in isolated hepatocytes after 20-min incubation without added substrate following addition of CH_3OH (25 mM after first addition, 38 mM after second addition). (B) Absorbance change after 10-min preincubation without substrate followed by 10 min with ethylmorphine (8 mM). (C) Absorbance change after 10-min preincubation with methionine (2 mM) followed by 10 min with ethylmorphine (8 mM)

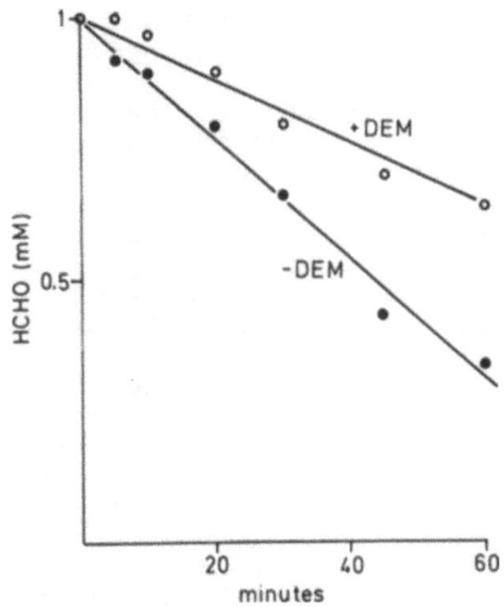

Fig. 4

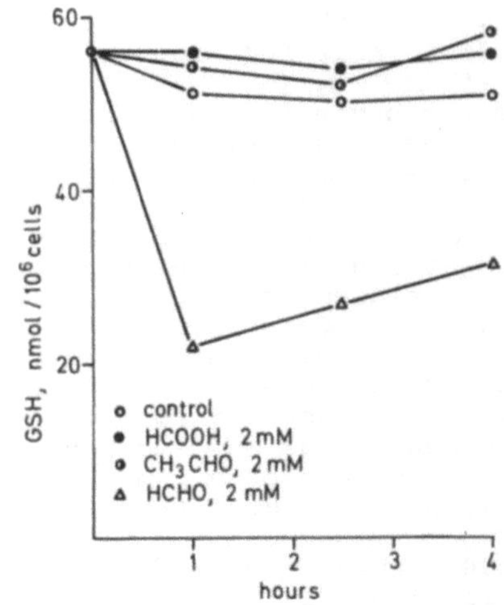

Fig. 5

Fig. 4. Effect of diethylmaleate pretreatment on formaldehyde loss in isolated hepatocytes. Cells were preincubated 3 min either without or with diethylmaleate (90 μM) and HCHO was added. HCHO concentration was subsequently assayed at indicated times

Fig. 5. GSH concentration in hepatocytes following addition of formaldehyde, formic acid and methanol. Assay of GSSG showed that the GSH was not converted to GSSG

dehyde dehydrogenase played a major role in oxidation of formaldehyde to formic acid. The loss of formaldehyde during incubation with hepatocytes is shown in Figure 4. Pretreatment of the cells with diethylmaleate decreased the rate by nearly 50%. Thus, the GSH-dependent process appears to have an important role in formaldehyde oxidation.

To examine this reaction pathway further we added formaldehyde to hepatocytes which had not been depleted of GSH. The effect of added formaldehyde on cellular GSH content is shown in Figure 5. GSH was rapidly lost, but unlike the experiments with added cytochrome P-450 substrate, GSH was not recovered as GSSG. Presumably the nonenzymatic reaction of GSH with formaldehyde results in conversion of GSH to S-hydroxymethylglutathione. Following this initial reaction, the level of GSH continually increases as the concentration of formaldehyde decreases. Thus, this enzyme system may be viewed as important in regeneration of GSH, a function which may be as important to the cell as oxidation of formaldehyde.

D. The Role of GSH in Formation of Conjugates with Electrophilic Drug Metabolites

The role of GSH in the formation of conjugates with electrophilic drug metabolites - most often formed by the cytochrome P-450-linked monooxygenase - is now well established. Thus, studies with a number of model compounds including halogenated benzenes and acetaminophen have shown that hepatotoxicity of such agents is preceded by GSH depletion and prevented under conditions of facilitated biosynthesis of this thiol (8,11).

During the last two years we have studied bromobenzene metabolism and toxicity in isolated hepatocytes in an attempt to clarify mechanisms underlying the hepatotoxic effect of electrophilic drug metabolites (21-23). In the case of bromobenzene, the 3,4-epoxide is most probably the reactive species which is formed predominantly by phenobarbital inducible cytochrome P-450 (8). Inactivation of this intermediate occurs primarily by conjugation with GSH and cytotoxicity is observed only when GSH has been depleted in the hepatocytes. In short-term experiments this may be achieved either by bromobenzene metabolism under conditions of inhibited GSH resynthesis, or by using hepatocytes with a lowered concentration of GSH produced by pretreatment of the rats with diethylmaleate prior to hepatocyte isolation (21,22).

Figure 6 shows that bromobenzene-induced cytotoxicity - measured as an increase in the permeability of the plasma membrane to NADH - is observed only with hepatocytes from rats pretreated with phenobarbital and DEM, and is preceded by GSH depletion. In hepatocytes from rats treated with phenobarbital, but not with DEM, there is a decrease, but no depletion, of GSH as a result of bromobenzene metabolism. This is due to resynthesis of GSH and is thus true only when the cells have access to cysteine and/or methionine in the medium. In the absence of these GSH precursors, bromobenzene exposure results in accelerated cell death also with hepatocytes with a normal GSH concentration at the beginning of incubation.

Figure 7 shows GSH levels in hepatocytes from phenobarbital (A) and phenobarbital and DEM-treated (B) rats incubated in the absence or presence of cysteine or methionine. In the former case, the presence of either amino acid prevents the decrease in GSH occurring upon incubation of the cells in a cysteine- and methionine-deficient medium (Fig.7A); with hepatocytes from DEM-treated rats there is a rapid resynthesis of cellular GSH (Fig.7B). Extracellular cysteine is taken up by the cells more rapidly than methionine (15,23) and, accordingly, GSH biosynthesis is faster with cysteine than with methionine in the medium. It should be noted, however, that cysteine oxidizes rapidly in the medium, and that cystine does not appear to enter the hepatocytes and cannot be utilized for replenishment of the GSH pool.

The concentration of intracellular cysteine in hepatocytes is low and the cells are depending on extracellular cysteine or methionine, which

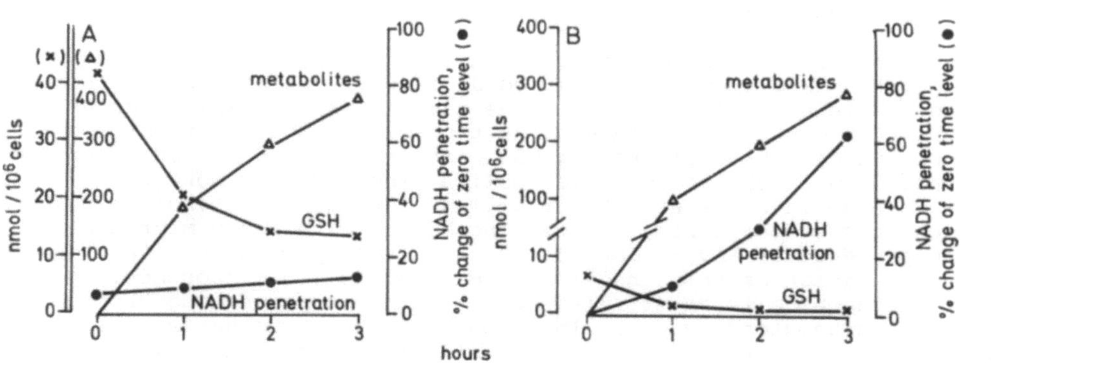

Fig. 6A and B. Bromobenzene metabolite formation, GSH concentration and NADH penetration in hepatocytes with and without pretreatment with diethylmaleate. (A) Cells from phenobarbital pretreated rat. (B) Cells from diethylmaleate and phenobarbital pretreated rat. [Data from (22)]

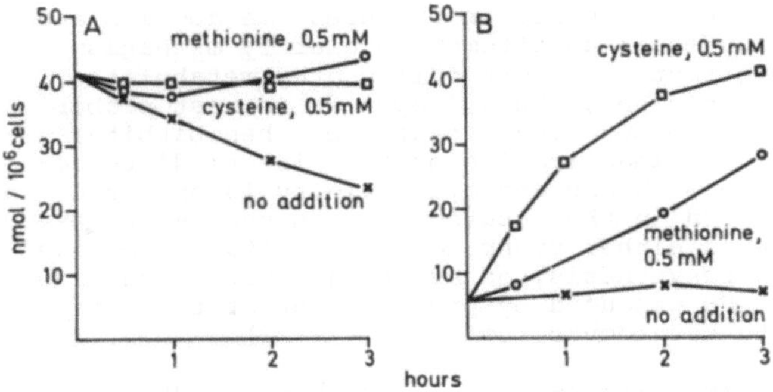

Fig. 7A and B. Effect of cysteine or methionine upon GSH concentration in hepatocytes in the presence of bromobenzene (0.6 mM). (A) Cells from phenobarbital pretreated rat. (B) Cells from diethylmaleate- and phenobarbital-pretreated rat. [Data from (23)]

Table 3. Effect of various amino acids on glutathione biosynthesis in hepatocytes isolated from phenobarbital and diethylmaleate-treated rats

Addition	GSH nmol/10^6 cells	
Incubation time:	0 h	2 h
Methionine, 0.5 mM	7.5	18.0
Methionine + serine, 0.5 mM	6.3	20.0
Methionine + serine + glycine, 0.2 mM + glutamate, 0.2 mM	6.2	19.2

yields cysteine by the cystathionine pathway, for GSH biosynthesis (15). On the other hand, the intracellular concentration of the other precursor amino acids - glycine, glutamate, and serine for production of cysteine via the cystathionine pathway - appears sufficient for optimal GSH biosynthesis during the first two to three hours of incubation under our conditions, since the addition of these amino acids to the medium did not markedly affect to rate of GSH synthesis observed with 0.5 mM methionine (Table 3).

Bromobenzene-induced cytotoxicity in hepatocytes is thus preceded by GSH depletion and prevented under conditions of facilitated GSH biosynthesis. This is illustrated by the data in Table 4 which further suggest that cysteine and methionine protect hepatocytes from bromobenzene toxicity by facilitating GSH biosynthesis and not by direct interaction with the reactive metabolite. Thus, GSH is more efficient than cysteine in decreasing the fraction of reactive metabolites of bromobenzene formed during aerobic incubation with liver microsomes in the presence of NADPH; methionine is virtually without effect on the accumulation of this fraction of metabolites in microsomes. In contrast, methionine is as efficient as cysteine in preventing bromobenzene toxicity in hepatocytes (22). In support of this hypothesis is also our recent observation that cysteine and methionine act by increasing the formation of GSH conjugates when added to rat hepatocytes incubated with acetaminophen (23).

Table 4. Effect of cysteine and methionine on formation of protein-bound metabolites of bromobenzene in liver microsomes and on GSH level and toxicity of bromobenzene in hepatocytes from phenobarbital- and diethylmaleate-treated rats [a]

Addition		Microsomes	Hepatocytes			
		Protein-bound metabolites, % of total metabolites	GSH nmol/10^6 cells		NADH penetration, % [b]	
Incubation time:		15 min	0	3 h	0	3 h
Bromobenzene,	0.6 mM	12.6	5.9	0.9	6	50
Bromobenzene + cysteine,	0.5 mM	3.9	6.1	8.7	8	12
Bromobenzene + methionine,	0.5 mM	10.6	6.7	10.2	8	11
Bromobenzene + GSH,	0.5 mM	1.4				

[a] Data compiled from (23).

[b] 100% level obtained by treatment of cells with 0.1% Triton X-100.

E. Conclusions

The results presented here demonstrate the role of glutathione in hepatic detoxication reactions in three different ways as summarized in Figure 8. In relation to its oxidative functions, the cytochrome P-450 monooxygenase system in isolated hepatocytes can generate H_2O_2, HCHO and epoxides, which are in turn detoxified by enzyme systems which utilize GSH either as a reductant, as a cofactor, or as a nucleo-

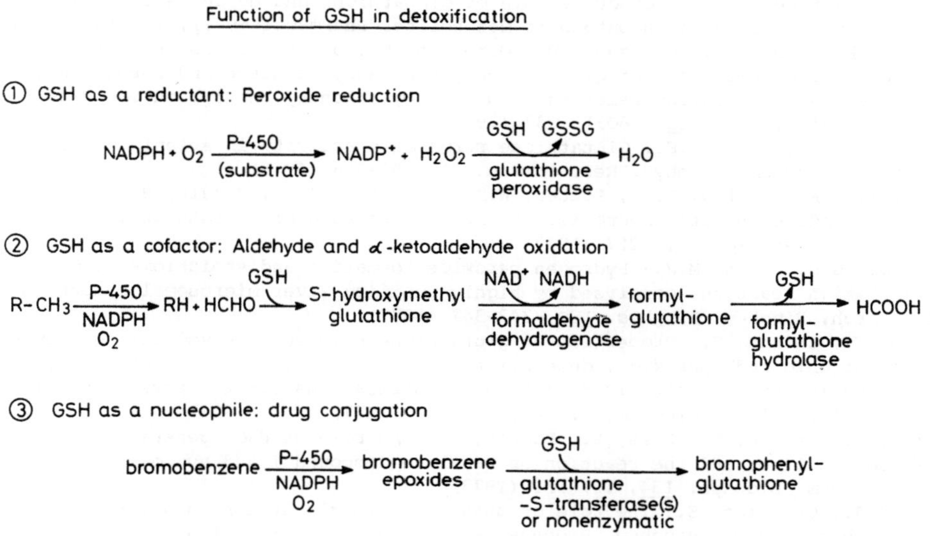

Fig. 8. Function of GSH in detoxification. Reaction types studied in this report characteristic of (1) reduction of peroxides, (2) oxidation of aldehydes and α-keto-aldehydes, and (3) conjugation of drugs

phile in conjugation reactions. Since intracellular accumulation of any of these reactive reaction products is associated with cell damage, the maintenance of a high intracellular concentration of GSH appears essential for prevention of cytotoxicity as a consequence of drug metabolism.

Acknowledgments. These studies were supported by a grant from the Swedish Medical Research Council (proj. O3X-2471).

References

1. Bartoli, G.M., Galeotti, T., Palombini, G., Parisi, G., Azzi, A.: Different contribution of rat liver microsomal pigments in the formation of superoxide anions and hydrogen peroxide during development. Arch. Biochem. Biophys. 184, 276-281 (1977)
2. Chance, B.: The effect of pH upon the equilibria of catalase compounds. J. Biol. Chem. 194, 483-496 (1952)
3. Flohé, L., Günzler, W.A.: Glutathione-dependent enzymatic oxidoreduction reactions. In: Glutathione: metabolism and function. Arias, I.M., Jakoby, W.B. (eds.), pp.17-34. New York: Raven Press 1976
4. Gillette, J.R., Brodie, B.B., LaDu, B.N.: The oxidation of drugs by liver microsomes: on the role of TPNH and oxygen. J. Pharmacol. Exp. Ther. 119, 532-540 (1957)
5. Goodman, J.I., Tephly, T.R.: A comparison of rat and human liver formaldehyde dehydrogenase. Biochim. Biophys. Acta 252, 489-505 (1971)
6. Hildebrandt, A.G., Speck, M., Roots, I.: Possible control of hydrogen peroxide production and degradation in microsomes during mixed function oxidation reaction. Biochem. Biophys. Res. Commun. 54, 968-975 (1973)
7. Jakoby, W.B., Habig, W.H., Keen, J.H., Ketley, J.N., Pabst, M.J.: Glutathione S-transferases: catalytic aspects. In : Glutathione: metabolism and function. Arias, I.M., Jakoby, W.B. (eds.), pp.189-202. New York: Raven Press 1976
8. Jollow, D.J., Mitchell, J.R., Zampaglione, N., Gillette, J.R.: Bromobenzene-induced liver necrosis. Protective role of glutathione and evidence for 3,4-bromobenzene oxide as the hepatotoxic metabolite. Pharmacology 11, 151-169 (1974)
9. Jones, D.P., Thor, H., Andersson, B., Orrenius, S.: Detoxification reactions in isolated hepatocytes. Role of glutathione peroxidase, catalase and formaldehyde dehydrogenase in reactions relating to N-demethylation by the cytochrome P-450 system. J. Biol. Chem. 253, 6031-6037 (1978)
10. Lawrence, R.A., Burk, R.F.: Glutathione peroxidase activity in selenium-deficient rat liver. Biochem. Biophys. Res. Commun. 71, 952-958 (1976)
11. Mitchell, J.R., Jollow, D.J., Potter, W.Z., Gillette, J.R., Brodie, B.B.: Acetaminophen-induced hepatic necrosis. IV. Protective role of glutathione. J. Pharmacol. Exp. Ther. 187, 211-217 (1973)
12. Nordblom, G.D., Coon, M.J.: Hydrogen peroxide formation and stoichiometry of hydroxylation reactions catalyzed by highly purified liver microsomal cytochrome P-450. Arch. Biochem. Biophys. 180, 343-347 (1977)
13. Oshino, N., Chance, B.: Properties of glutathione release observed during reduction of organic hydroperoxide, demethylation of aminopyrine and oxidation of some substances in perfused rat liver, and their implications for the physiological function of catalase. Biochem. J. 162, 509-525 (1977)
14. Oshino, N., Chance, B., Sies, H., Bücher, T.: The role of H_2O_2 generation in perfused rat liver and the reaction of catalase Compound I and hydrogen donors. Arch. Biochem. Biophys. 154, 117-131 (1973)
15. Reed, D.J., Orrenius, S.: The role of methionine in glutathione biosynthesis by isolated hepatocytes. Biochem. Biophys. Res. Commun. 77, 1257-1264 (1977)
16. Richter, C., Azzi, A., Weser, U., Wendel, A.: Hepatic microsomal dealkylations. Inhibition by a tyrosine copper (II) complex provided with superoxide dismutase activity. J. Biol. Chem. 252, 5061-5066 (1977)

17. Sies, H., Bartoli, G.M., Burk, R.F., Waydhas, C.: Glutathione efflux from perfused rat liver after phenobarbital-treatment, during drug oxidations, and in selenium deficiency. Eur. J. Biochem. 89, 113-118 (1978)
18. Sies, H., Bücher, T., Oshino, N., Chance, B.: Heme occupancy of catalase in hemoglobin-free perfused rat liver and of isolated rat liver catalase. Arch. Biochem. Biophys. 154, 106-116 (1973)
19. Sies, H., Chance, B.: The steady-state level of catalase Compound I in isolated hemoglobin-free perfused rat liver. FEBS Lett. 11, 172-176 (1970)
20. Strittmatter, P., Ball, E.G.: Formaldehyde dehydrogenase, a glutathione-dependent enzyme system. J. Biol. Chem. 213, 445-461 (1955)
21. Thor, H., Moldéus, P., Kristoferson, A., Högberg, J., Reed, D.J., Orrenius, S.: Metabolic activation and hepatotoxicity. Metabolism of bromobenzene in isolated hepatocytes. Arch. Biochem. Biophys. 188, 114-121 (1978)
22. Thor, H., Moldéus, P., Hermanson, R., Högberg, J., Reed, D.J., Orrenius, S.: Metabolic activation and hepatotoxicity. Toxicity of bromobenzene in hepatocytes isolated from phenobarbital- and diethylmaleate-treated rats. Arch. Biochem. Biophys. 188, 122-129 (1978)
23. Thor, H., Moldéus, P., Orrenius, S.: Metabolic activation and hepatotoxicity. Effect of cysteine, N-acetylcysteine and methionine on glutathione biosynthesis and bromobenzene toxicity in isolated rat hepatocytes. Arch. Biochem. Biophys., in press (1979)
24. Thurman, R.G., Ley, H.G., Scholz, R.: Hepatic microsomal ethanol oxidation. Hydrogen peroxide formation and the role of catalase. Eur. J. Biochem. 25, 420-430 (1972)
25. Uotila, L., Koivusalo, M.: Formaldehyde dehydrogenase from human liver. Purification, properties and evidence for the formation of glutathione thiol esters by the enzyme. J. Biol. Chem. 249, 7653-7663 (1974)
26. Uotila, L., Koivusalo, M.: Purification and properties of S-formylglutathione hydrolase from human liver. J. Biol. Chem. 249, 7664-7672 (1974)
27. Waydhas, C., Weigl, K., Sies, H.: The disposition of formaldehyde and formate arising from drug N-demethylations dependent on cytochrome P-450 in hepatocytes and in perfused rat liver. Eur. J. Biochem. 89, 143-150 (1978)

Lipid Peroxidation Induced by Ethanol and Halogenated Hydrocarbons in vivo as Measured by Ethane Exhalation

H. KAPPUS, U. KÖSTER, D. KÖSTER-ALBRECHT, H. KIECZKA, and H. REMMER

Summary

The measurement of ethane and propane expiration of rats in vivo after treatment with foreign compounds is described. Carbon tetrachloride (CCl_4) and bromotrichloromethane induce ethane expiration of rats depending on the dose of the chemical. Furthermore, control animals also expire some ethane during the 180 min of the experiment.

Trichloroethylene, vinyl chloride, halothane and ethanol are also able to induce ethane production by the animals in vivo.

The amount of ethane expired by the animals when treated with CCl_4 depends on the oxygen concentration in the atmosphere, higher oxygen concentrations leading to lower amounts of ethane. The same was found for lipid peroxidation in rat liver microsomes where a maximum of the malondialdehyde formation induced by CCl_4 was found in presence of 5% -7% oxygen, whereas higher oxygen concentrations decreased the malondialdehyde formation.

The results are discussed in relation to the hepatotoxicity of halogenated hydrocarbons and of ethanol. The measurement of alkanes occurring during lipid peroxidation in vivo is recommended for studies on lipid peroxidation induced by foreign compounds including the protective role of glutathione in this toxic process.

A. Introduction

The peroxidation of membrane lipids is one of the molecular mechanisms which might be responsible for the toxicity induced by foreign compounds taken up by the organism (Plaa and Witschi, 1976). During the last 30 years numerous reports appeared which dealt with the role of glutathione in diseases. For example, it is well known that glutathione peroxidase metabolizes lipid peroxides potentially toxic to normal cell constituents (Flohé et al., 1976). Furthermore, sulfhydryl compounds, like glutathione in vivo, react in vitro with lipid hydroperoxides (Gardner et al., 1976, 1977). Therefore, studies on the interrelationship between the lipid peroxide forming and the lipid peroxide inactivating capacities of various cells are in progress.

Several methods are available to measure lipid peroxidation in vitro (Plaa and Witschi, 1976). However, only a few methods are also applicable in vivo. On the other hand, none of the methods applied discriminates between different products occurring during different steps of the lipid peroxidation process. Only the measurement of increased conjugated dienes of lipids seems to represent a measure of the first attack of a free radical on the unsaturated lipid molecule (Recknagel and Glende, 1973; Plaa and Witschi, 1976). However, this method is discontinuous, time-consuming and relatively insensitive.

The measurement of gaseous alkanes formed probably by the breakage of unsaturated fatty acids attacked by a free radical molecule might be a sensitive and rapid method for the determination of the first reaction step occurring during lipid peroxidation. The auto-oxidation of unsaturated fatty acids as well as the lipo-oxygenase reaction with those fatty acids produce several alkanes and alkenes, the nature of the gaseous compounds being dependent on the type of fatty acids (Lieberman and Mapson, 1964; Dumelin and Tappel, 1977). It is now well established that under various biological conditions, such as vitamin E or selenium deficiencies, alkanes, especially ethane and pentane, are expired by animals in vivo (Dillard et al., 1977; Hafeman and Hoekstra, 1977b). Under these conditions lipid peroxidation does presumably occur. Ethane and ethylene are also formed by plants due to aerobic lipid degradation (Konze and Elstner, 1978).

Furthermore, Riely et al. (1974) were the first to demonstrate that ethane is expired by animals after treatment with CCl_4. Ethane was also formed by liver homogenates incubated in the presence of CCl_4. The results were confirmed by several authors including our own work (Hafeman and Hoekstra, 1977a; Köster et al., 1977; Lindstrom and Anders, 1978). We extended the in vivo experiments of Riely et al. (1974) to the measurement of propane (Köster et al., 1977).

B. Experiments

The different laboratories engaged in the determination of alkanes by animals under various conditions use different sampling methods. We use an all-glass system which is flushed with purest air or oxygen before the first measurement of the alkanes. The experimental design as well as the animals used were the same as described previously (Köster et al., 1977).

Figures 1 and 2 demonstrate that ethane expiration by the animals is dependent on the dose of CCl_4 or $CBrCl_3$, respectively, given to the animals before they were placed into the closed system. The dose response curves were obtained from time curves measured as described previously (Köster et al., 1977). During the present experiments the ethane expiration induced by CCl_4 (1 g/kg) was low compared to the amounts measured during the previous experiments (Köster et al., 1977). However, we know that there exists a relatively high variation of the results when different animals were used. The absolute amount of ethane formed is probably dependent on some unknown factors, like the antioxidant content of food, etc.

Figure 1 demonstrates that the ethane expiration of the animals increases with increasing doses of CCl_4, up to 1.5 g/kg, whereas the amount of ethane exhaled by the animals after $CBrCl_3$-doses higher than 140 mg/kg remains constant (Fig.2). Doses higher than 330 mg/kg lead to a further increase in the ethane formation. However, all animals dosed with such high $CBrCl_3$-doses died during or after the 180-min period of the experiment. Therefore, the interpretation of this high ethane production rate of these animals is relatively difficult.

Table 1 summarizes our ethane data obtained after treatment of rats with different halogenated aliphatic hydrocarbons and with ethanol. All compounds examined are able to induce ethane expiration of the animals. Propane was also formed to some extent by the animals treated with either of these compounds (data not shown). Table 1 also demonstrates that control animals exhale low amounts of ethane. This increase in ethane expiration with time does occur without any pretreatment of the animals.

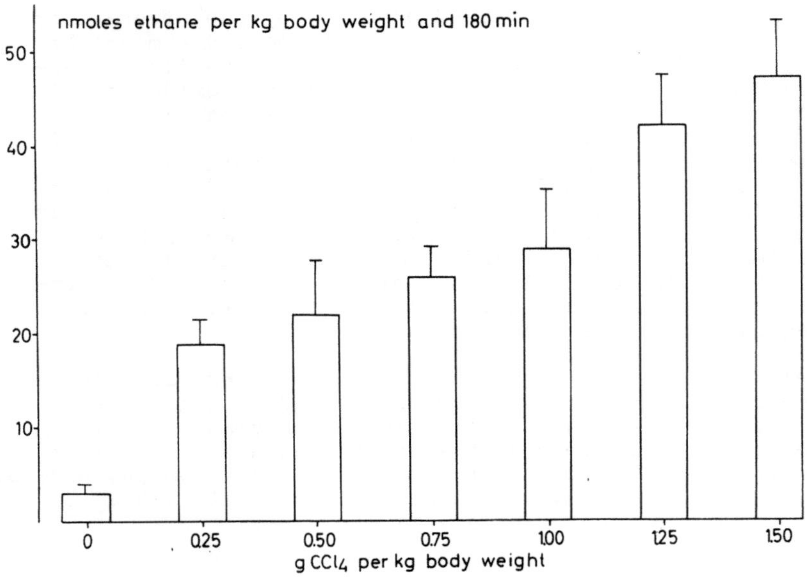

Fig. 1. Dose response curve of the ethane expiration of rats on the concentration of CCl_4 (in olive oil) given i.p. The rats were treated at zero time and then the ethane expiration was checked every 30 min and followed for 180 min. Mean values ±s.d. of four different experiments with different rats are shown

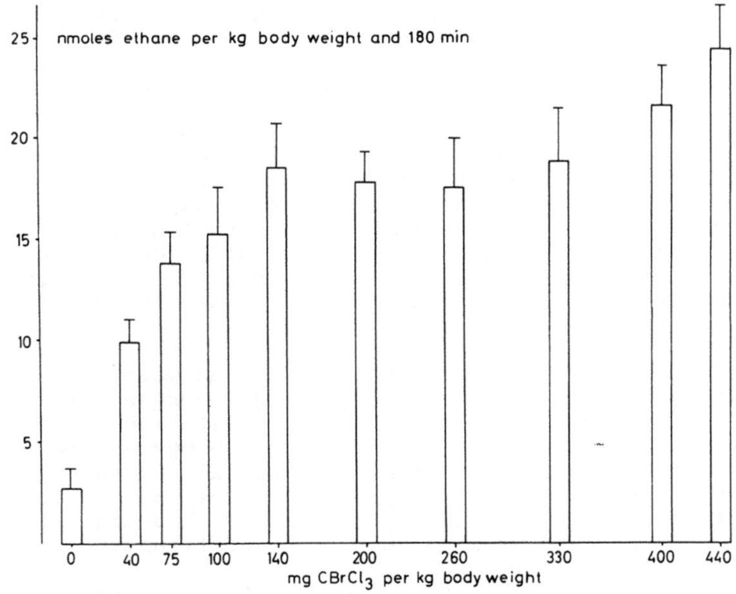

Fig. 2. Dose response curve of the ethane expiration of rats on the concentration of $CBrCl_3$ (in olive oil) given i.p. The rats were treated at zero time and then the ethane was checked every 30 min and followed for 180 min. Mean values ±s.d. of four different experiments with different rats are shown

Table 1. Ethane expiration of rats during exposition or after treatment with foreign compounds. The controls were untreated. The foreign compounds were either injected i.p., dissolved in water or olive oil, or were initially present in the atmosphere as gaseous compounds. The ethane values are means of at least four different experiments with different rats obtained from time curves after 180 min

	nmol ethane / kg body wt. and 180 min
Controls	2.7
Trichloroethylene (2%)	5.7
Vinyl Chloride (2%)	7.9
Halothane (2%)	9.5
Ethanol (5 g/kg)	12.5
$CBrCl_3$ (250 mg/kg)	17.8
CCl_4 (250 mg/kg)	18.7

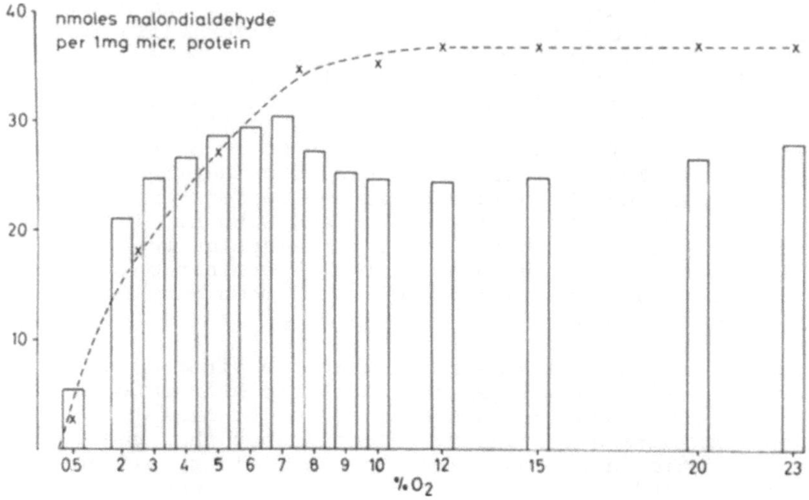

Fig. 3. Malondialdehyde formation in rat liver microsomes incubated in presence of CCl_4 (1 µl/ml) or Fe^{2+}-ADP and different oxygen concentrations. The content of the incubation mixtures was the same as described previously (Kappus et al.,1977). Each open column represents a single incubation performed in the presence of CCl_4 for 30 min with identical microsomal suspensions. *Each cross* (x--x--x--x) represents a single incubation performed in presence of Fe^{2+}-ADP for 30 min with identical microsomal suspensions. The *abscissa* shows the oxygen concentrations present above the incubation mixtures equilibrated before for 10 min

Some preliminary studies, not shown, indicate that the CCl_4-induced ethane expiration of the rats depends on the oxygen concentration present in the atmosphere. All the experiments described above were performed in an 100% oxygen atmosphere. When we decreased the oxygen concentration in the atmosphere to about 20%, the amount of ethane expired by the animals almost doubled. This effect was observed only in CCl_4-treated animals, but not in controls (data not shown). Because we were interested whether this oxygen effect is due to the difference in the enzymatic activation of CCl_4 to free radicals in presence of different oxygen concentrations we carried out some experiments with rat liver microsomes using NADPH and CCl_4 in a closed system. Figure 3 demonstrates that CCl_4-induced lipid peroxidation in microsomes, measured by the malondialdehyde estimation, is negligible at low oxygen concentrations and reaches its maximal value at about 5%-7% oxygen present in the atmosphere. Higher oxygen concentrations lead to lower amounts of malondialdehyde. This oxygen dependence curve was repeatedly found with rat liver microsomes. It was obtained by flushing the desired atmosphere through the microsomal suspension for 10 min before the addition of CCl_4. The microsomes were then incubated under the same atmosphere in presence of CCl_4. On the other hand, Figure 3 demonstrates that the Fe^{2+}-ADP-induced lipid peroxidation shows a simple oxygen dependence, with a maximum at about 10% oxygen, as already described by others (Lumper et al., 1968).

C. Discussion

It is well known that several aliphatic halogenated hydrocarbons, as well as ethanol, damage the hepatocytes in vivo. However, the biochemical mechanism leading to necrosis and cirrhosis of liver cells is still unknown. During this injuring process a protective role is attributed to glutathione in the hepatocytes. A relationship between the glutathione content of the hepatocytes and the cell damages induced by CCl_4 or ethanol has been assumed (Plaa and Witschi, 1976). The mechanism of the protective effect of glutathione on CCl_4- or ethanol-induced hepatotoxicity is however unclear. Lipid peroxidation is one of the proposed mechanisms which could explain the lesions of cells produced by foreign compounds (Mead, 1976; Plaa and Witschi, 1976). However, the processes occurring during lipid peroxidation remain unknown, probably because suitable methods for the determination of free radical attacks on unsaturated fatty acids of lipid membranes in vivo were lacking until recent years.

Therefore, we used the new method measuring alkanes expired by animals in vivo which were treated with halogenated aliphatic hydrocarbons or with ethanol. All the compounds tested are suspected to cause liver damage by a lipid peroxidation mechanism. We could show that the ethane and also propane expiration are closely related to lipid peroxidation in vivo (Köster et al., 1977), although there exist some uncertainties concerning the chemical mechanism of this alkane production. Such a close relationship was also found by others (Riely et al., 1974; Hafeman and Hoekstra, 1977a; Lindstrom and Anders, 1978; Litov et al., 1978). From the previous work of Lieberman and Mapson (1964) it has been concluded that ethane is formed from linolenic acid (ω-3 fatty acid) and pentane from linoleic acid or arachidonic acid (ω-6 fatty acids). However, these proposed reaction schemes cannot explain the formation of propane and of other hydrocarbons which occur during lipid peroxidation. An additional proof that alkanes are formed by a lipid peroxidation reaction is from the results obtained by Dumelin et al. (1978) who measured ethane and pentane after ozone exposure of animals, and the results of Dougherty and Hoekstra (1977) who obtained increased amounts of ethane after intoxication of animals with iron ions.

Our preliminary results on the oxygen dependence of ethane expiration after treating animals with CCl_4 (see experiments), compared to our in vitro results on malondialdehyde formation under different oxygen concentrations (Fig.3), and the results of Nastainczyk et al. (1978) who measured an increase in halothane-cytochrome P-450 (red.)-complex formation in microsomes and in the whole liver, indicate that lipid peroxidation might be involved in the toxic process induced by these compounds in the liver. It might depend on the reductive metabolism of the halogenated hydrocarbons which increases under reduced oxygen pressure (Nastainczyk et al., 1978; Reiner et al., 1972) leading to more free radicals. This is followed by a higher formation rate of lipid peroxides (Recknagel and Glende, 1973; Reiner et al., 1972). A reduced oxygen pressure has also been found in the centrolobular hepatocytes where liver necrosis starts.

The increase in the formation of lipid peroxides could overload the peroxide inactivating capacities of the glutathione peroxidase by depleting glutathione of the hepatocytes which is released from the cells as oxidized glutathione (Sies and Summer, 1975; Bartoli and Sies, 1978; Burk et al., 1978). The increased amount of lipid peroxides could then damage the cell (Mead, 1976).

From our studies and the studies of others it can be concluded that the measurement of alkanes occurring after intoxication with foreign compounds is the method of choice to elucidate the role of glutathione in the toxic process induced by foreign chemicals in the hepatocytes, if these compounds are able to catalyze a peroxidation of lipid membranes. Therefore, we have to perform studies on the measurement of both the lipid peroxidation process and the glutathione content of the hepatocytes simultaneously. This will be done in the near future.

Acknowledgments. This study has been supported by the Deutsche Forschungsgemeinschaft, Bonn Bad-Godesberg, Germany. The skilful technical assistance of Mrs. E. Steinhilber during the in vitro experiments is gratefully acknowledged.

References

Bartoli, G.M., Sies, H.: Reduced and oxidized glutathione efflux from liver. FEBS Lett. 86, 89-91 (1978)

Burk, R.F., Nishiki, K., Lawrence, R.A., Chance, B.: Peroxide removal by selenium-dependent and selenium-independent glutathione peroxidases in hemoglobin-free perfused rat liver. J. Biol. Chem. 253, 43-46 (1978)

Dillard, C.J., Dumelin, E.E., Tappel, A.L.: Effect of dietary vitamin E on expiration of pentane and ethane by the rat. Lipids 12, 109-114 (1977)

Dougherty, J.J., Hoekstra, W.G.: Iron as a stimulator of lipid peroxidation in vivo in selenium and vitamin E deficient rats. Federation Proc. 36, 1151 (1977)

Dumelin, E.E., Dillard, C.J., Tappel, A.L.: Breath ethane and pentane as measures of vitamin E protection of Macaca radiata against 90 days of exposure to ozone. Environm. Res. 15, 38-43 (1978)

Dumelin, E.E., Tappel, A.L.: Hydrocarbon gases produced during in vitro peroxidation of polyunsaturated fatty acids and decomposition of preformed hydroperoxides. Lipids 12, 894-900 (1977)

Flohê, L., Günzler, W.A., Ladenstein, R.: Glutathione peroxidase. In: Glutathione: metabolism and function. Arias, I.M., Jakoby, W.B. (eds.), pp.115-135. New York: Raven Press 1976

Gardner, H.W., Kleiman, R., Weisleder, D., Inglett, G.E.: Cysteine adds to lipid hydroperoxide. Lipids 12, 655-660 (1977)

Gardner, H.W., Weisleder, D., Kleiman, R.: Addition of N-acetylcysteine to linoleic acid hydroperoxide. Lipids 11, 127-134 (1976)

Hafeman, D.G., Hoekstra, W.G.: Protection against carbon tetrachloride-induced lipid peroxidation in the rat by dietary vitamin E, selenium, and methionine as measured by ethane evolution. J. Nutr. 107, 656-665 (1977a)

Hafeman, D.G., Hoekstra, W.G.: Lipid peroxidation in vivo during vitamin E and selenium deficiency in the rat as monitored by ethane evolution. J. Nutr. 107, 666-672 (1977b)

Kappus, H., Kieczka, H., Scheulen, M., Remmer, H.: Molecular aspects of catechol and pyrogallol inhibition of liver microsomal lipid peroxidation stimulated by ferrous ion-ADP-complexes or by carbon tetrachloride. Naunyn-Schmiedeberg's Arch. Pharmacol. 300, 179-187 (1977)

Konze, J.R., Elstner, E.F.: Ethane and ethylene formation by mitochondria as indication of aerobic lipid degradation in response to wounding of plant tissue. Biochim. Biophys. Acta 528, 213-221 (1978)

Köster, U., Albrecht, D., Kappus, H.: Evidence for carbon tetrachloride- and ethanol-induced lipid peroxidation in vivo demonstrated by ethane production in mice and rats. Toxicol. Appl. Pharmacol. 41, 639-648 (1977)

Lieberman, M., Mapson, L.W.: Genesis and biogenesis of ethylene. Nature (London) 204, 343-345 (1964)

Lindstrom, T.D., Anders, W.: Effect of agents known to alter carbon tetrachloride hepatotoxicity and cytochrome P-450 levels on carbon tetrachloride-stimulated lipid peroxidation and ethane expiration in the intact rat. Biochem. Pharmacol. 27, 563-567 (1978)

Litov, R.E., Dillard, C.J., Downey, J., Irving, D., Sagai, M., Gee, D., Tappel. A.L.: Measurement of in vivo lipid peroxidation via pentane: effect of vitamin E, selenium and polyunsaturated fats and assay for acute toxic reactions. Federation Proc. 37, 706 (1978)

Lumper, L., Plock, H.J., Staudinger, H.: Untersuchungen zur Abhängigkeit der Lipid-peroxydation in Rattenlebermikrosomen vom Sauerstoffpartialdruck. Hoppe-Seyler's Z. Physiol. Chem. 349, 1185-1190 (1968)

Mead, J.F.: Free radical mechanisms of lipid damage and consequences for cellular membranes. In: Free radicals in biology. Pryor, W.A. (ed.), pp.51-68. New York-San Francisco-London: Academic Press 1976

Nastainczyk, W., Ullrich, V., Sies, H.: Effect of oxygen concentration on the reaction of halothane with cytochrome P-450 in liver microsomes and isolated perfused rat liver. Biochem. Pharmacol. 27, 387-392 (1978)

Plaa, G.L., Witschi, H.: Chemicals, drugs, and lipid peroxidation. Ann. Rev. Pharmacol. Toxicol. 16, 125-141 (1976)

Recknagel, R.O., Glende, E.A.: Carbon tetrachloride hepatotoxicity: an example of lethal cleavage. CRC Crit. Rev. Toxicol. 2, 263-297 (1973)

Reiner, O., Athanassopoulos, S., Hellmer, K.H., Murray, R.E., Uehleke, H.: Bildung von Chloroform aus Tetrachlorkohlenstoff in Lebermikrosomen. Lipidperoxidation und Zerstörung von Cytochrom P-450. Arch. Toxicol. 29, 219-233 (1972)

Riely, C.A., Cohen, G., Lieberman, M.: Ethane evolution: a new index of lipid peroxidation. Science 183, 208-210 (1974)

Sies, H., Summer, K.H.: Hydroperoxide-metabolizing systems in rat liver. Eur. J. Biochem. 57, 503-512 (1975)

Drug-induced Lipid Peroxidation in Mouse Liver

A. WENDEL, S. FEUERSTEIN, and K.-H. KONZ

Summary

It was demonstrated that acute intoxication of mice leads to in vivo formation of lipid peroxides following ethane exhalation of the animals. Fed mice exhalated 5 nmol of ethane/kg/h after 500 mg/kg i.p. of acetaminophen, while total liver GSH dropped to 50%. 2 days of starvation led to a 40% decrease of hepatic GSH; with the same dose these animals exhalated 125-150 nmol ethane/kg/h with a final drop to 10% residual GSH. In vivo expired ethane and in vitro determined malondialdehyde correlated fairly well. No pentane was detectable. Refeeding of acetaminophen-treated starved mice resulted in a recovery of liver GSH to 60% of the normal value. Untreated starved mice reached hepatic GSH levels of 400% above normal within 16 h.

After treatment with diethylmaleate, the same drastic GSH depletion occurred as with acetaminophen, but no significant ethane evolution was observed. After an additional load of 500 mg/kg acetaminophen the animals expired 25-30 nmol of ethane/kg/h.

These findings indicate that GSH depletion alone does not lead to lipid peroxidation in vivo. It is suggested that one part of liver GSH is immediately available for biotransformation and another pool which is sensitive to starvation, affords long-term protection in a different way.

A. Introduction

Lipid peroxidation in vivo seems to be a basic deteriorative reaction in the chain of events which lead to liver injury (11,14). It is generally agreed that intrahepatic glutathione is able to afford protection against liver dysfunction by at least two ways. Firstly as a substrate of glutathione peroxidase, GSH serves to reduce a large variety of hydroperoxides before they attack unsaturated lipids or converts already formed lipid hydroperoxides to the corresponding hydroxy compounds (3). Secondly as a substrate of glutathione-S-transferases, it enables the liver to detoxify many foreign compounds or their metabolites and to excrete the product, preferably into the bile (2). Both ways lead to considerable losses of liver glutathione, either in the form of GSSG (13) or in the form of a conjugate with an electrophilic compound. In addition, the alimentary supply of the liver was shown to influence markedly the glutathione level of the organ (15), presumably via the rate of degradation by γ-glutamyltransferase (6). Despite investigation of details in vitro the quantitative contribution of total liver glutathione to the antiperoxidative protection potential in vivo of the organ is not well understood.

The aim of the present study was to find out whether acute acetaminophen intoxication leads to in vivo lipid peroxidation. Of special interest was the minimal threshold of intrahepatic glutathione which prevents peroxidation, and its relation to the nutritional status of the organism.

B. Materials and Methods

Female albino mice weighing an average of 15 g were raised on a diet (C 1018, Altromin. Lage W.Germany) which was low in vitamin E and not supplemented with additional selenium; palmitin as the fat source was replaced by soybean oil (4%).

The animals were injected i.p. with a 0.25 M solution of acetaminophen (10) in 0.9% saline. Alternatively, diethylmaleate was applied i.p. as a 20% solution in sesame oil. Ethane evolution was monitored essentially as described by Riely et al. (12). Single animals breathed in a 800 ml desiccator; CO_2 was absorbed by 80 g "Atemkalk" (Rhein-Pharma Arzneimittelwerk Plankstadt) and the equivalent amount of gas was replaced by O_2 from a reservoir. After mixing, 5 ml of air were removed with a syringe and analyzed in a Hewlett-Packard Model 5750 G gas chromatograph with FID detection. N_2 was used as carrier gas, the column (0.4 x 400 cm) was filled with Porapak Q-100-120 mesh (Applied Science Lab.Inc.). Analysis was run at the highest sensitivity range isothermally at 70°C. The system was calibrated (including desiccator) with a 0.73 parts/10^6 ethane/N_2 standard supplied by Messer-Grießheim, Duisburg. A mixture of ethane, propane, and pentane supplied by the same company was used to calibrate and control the analytical system.

For in vitro determination of lipid peroxidation, similarly treated groups of animals were killed. The livers and kidneys were removed and homogenized in ice-cold 3% metaphosphoric acid. The supernatant of a 1 min centrifugation in an Eppendorf centrifuge at full speed was used for the determination of total glutathione (1) and malondialdehyde (5). Protein was determined in the homogenate (9).

C. Results

Acute acetaminophen intoxication of mice led to depletion of liver glutathione within 5 h. At a dose of 500 mg/kg acetaminophen a residual glutathione level of 30 nmol/mg protein was measured, i.e., 50% of the tripeptide was eliminated from the liver. In contrast to what we expected these animals did not expire significant amounts of ethane. Taking advantage of the observations made by Tateishi et al. (15) we investigated the effects of starvation on the acetaminophen-induced glutathione depletion of the liver.

Figure 1 shows that starving the animals reduced the hepatic glutathione content within one day to 60% compared to unstarved controls. Prolonged starvation did not result in a further decrease beyond this level. If the animals were injected in this state with acetaminophen the hepatic glutathione decreased to 11% of controls within the first hour. Upon refeeding this level rose again to 36%-77% in the treated animals whereas untreated mice reached 350%-450% above normal.

As shown in Figure 2, the starved animals treated with acetaminophen produced up to 150 nmol of ethane/kg/h, whereas unstarved treated animals expired 5 nmol/kg/h. Above 300 mg/kg the mortality ranged around 20%. Below 375 mg/kg no measurable ethane evolution occurred.

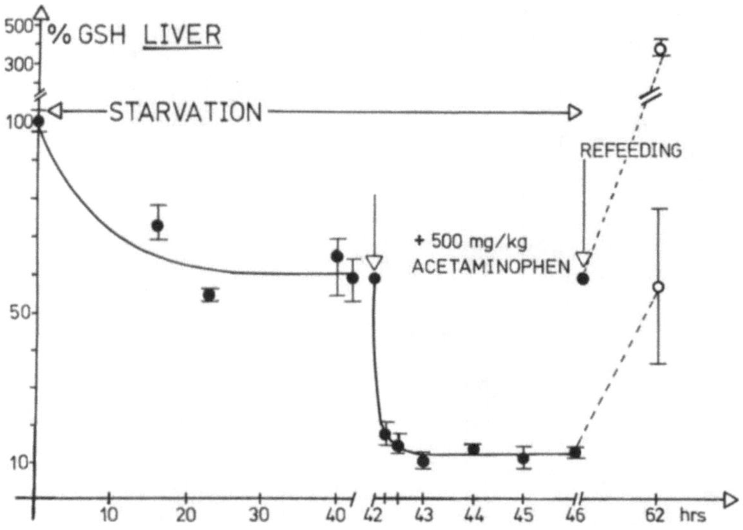

Fig. 1. Total glutathione content of mouse livers during starvation, subsequent treatment with acetaminophen, and refeeding. The values represent means of 3 animals and their standard deviation, respectively. 100% corresponds to 61 nmol of glutathione/mg protein

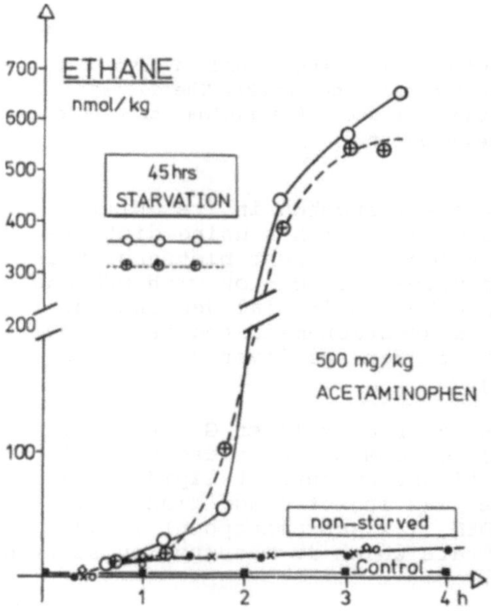

Fig. 2. Ethane exhalation of mice treated with 500 mg/kg acetaminophen i.p. after starvation for 2 days. Unstarved and untreated animals served as controls (n=5). The *open circles* and *crossed open circles* represent two individuals

The amount of expired ethane and the residual glutathione level as determined post mortem in the livers are inversely proportional (Fig.3). The malondialdehyde formation in these livers correlated with the ethane evolution in vivo with a correlation coefficient of 0.74 (data not shown).

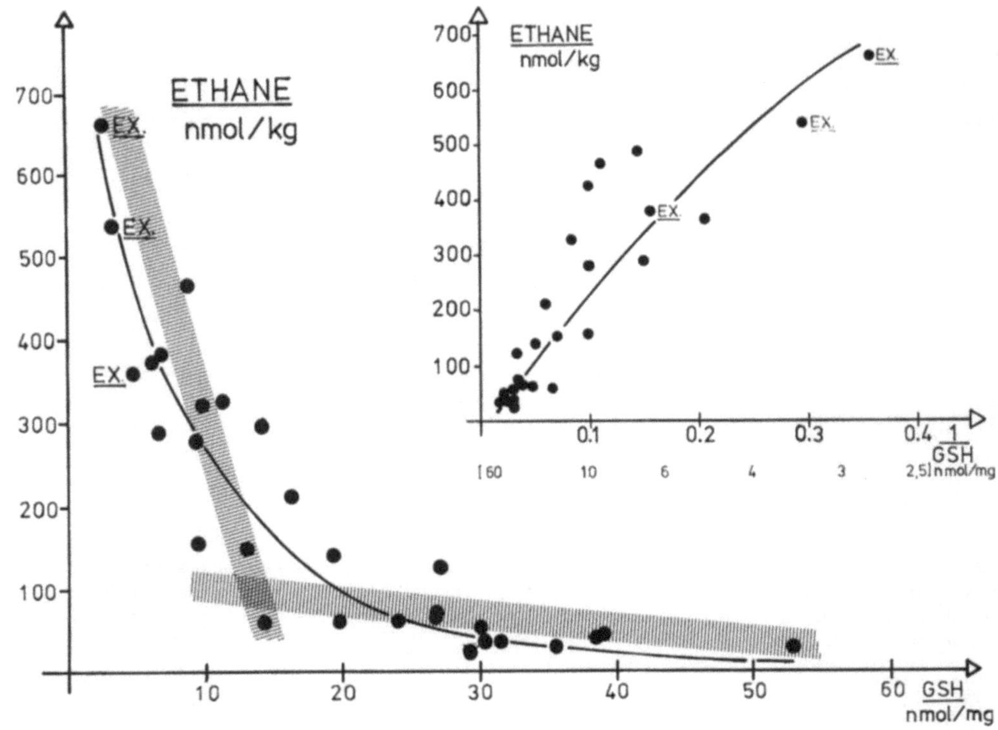

Fig. 3. Quantitative relationship between in vivo expired ethane and in vivo post mortem determined hepatic total glutathione in starved mice (n=29). The *dashed straight lines* show the extrema expected for the existence of a minimal threshold of liver glutathione. Ex = exitus during the experiment

In order to clarify the specificity of acetaminophen in its ability to induce lipid peroxidation, a series of experiments using diethylmaleate (DEM) as a substrate for glutathione-dependent biotransformation (2) was performed. In addition, double-intoxication with DEM and acetaminophen should allow to decide whether an initial decrease of glutathione per se or the specifically starvation-mediated depletion of glutathione prior to drug treatment, render the liver vulnerable for the deleterious action of the drug.

Table 1 shows that a double dose of DEM decreases liver GSH to 17%. Starved mice treated with a triple dose of DEM showed a decrease of GSH to only 4%. In both cases no significant amounts of lipid peroxides were produced unless acetaminophen was injected additionally. In these triple-treated animals (2 x DEM, 1 x acetaminophen) lipid peroxidation amounts to one-fifth the rate which is measured in starved mice treated with acetaminophen alone.

D. Discussion

The present data demonstrate that indeed highly-dosed acetaminophen administration leads to in vivo formation of lipid peroxides, but only under certain circumstances. From their studies on the regulation of

Table 1. Effect of repeated i.p. treatment of mice on hepatic glutathione, ethane exhalation in vivo and MDA formation in vitro with DEM and acetaminophen. The values are arithmetic means

Drug dose	Starved (h)	Total liver GSH nmol/mg	Ethane nmol/kg/h	Liver MDA pmol/mg	No. of animals
3 x DEM [a] 400 mg/kg	48	2.3	3	<10	n=3
2 x DEM [b] 400 mg/kg	not	10.4	3	44	n=3
2 x DEM [c] 400 mg/kg 1 x Acetaminophen 500 mg/kg	not	2.4	30	49	n=4

[a] Injection at 0, 20, 40 min of the experiment.

[b] Same at 0, 20 min.

[c] Same at 0, 20 (DEM), 30 (Acetaminophen) min.

the GSH level in rat liver Tateishi et al. (15) postulated the existence of a high- and a low-turnover pool in this organ with half-lives of 1.7 and 28 h, respectively. It is tempting to assume that we deal in this investigation also with two pools with different functions: the high-turnover pool may be immediately available for biotranformation in emergency situations, the low-turnover pool may provide a high degree of protection against lipid peroxidation (cf. Figs.1 and 2). Generally in line with this idea is the observation that reduction of intrahepatic GSH by 83% with DEM results in only one-fifth of the ethane production that occurs with the same dose of acetaminophen after a GSH decrease by 40% following starvation. This means that a minimal threshold of residual glutathione resides below 10% of the normal level. Keeping in mind that GSH and GSSG were not analytically differentiated and that contamination of the liver homogenate with blood GSH is likely, the true minimal GSH requirement is very low.

The critical dose of acetaminophen seems to be above 375 mg/kg in the system used here. It is interesting to note that Mitchell and coworkers (7) observed above this dose a sharp increase of the protein binding of the drug's metabolite, which is held responsible for the degree of lesion. This and the lacking ability of DEM to induce lipid peroxidation favor the idea of a dual role of acetaminophen, initial depletion of GSH and a surplus of reactive metabolite, to lead to severe liver injury.

Although widely used, malondialdehyde formation has never been fully accepted as an appropriate parameter for the measurement of lipid peroxidation. Confirming the conclusions of Riely et al. (12) and others (4), this investigation shows that ethane exhalation seems to fulfill the criteria for measurement of such a complex reaction with a 10-fold increased sensitivity compared to the malondialdehyde method. No pentane, as reported by others (8), was detected.

As long as additional consequences of fasting for the animal's metabolism are not investigated, it seems premature to attribute decreased hepatic resistance against lipid peroxidation solely to the starvation-sensitive liver GSH. Although our knowledge of the molecular basis is still poor, the drug-induced lipid peroxidation in starved mice as measured via ethane exhalation offers the possibility to screen in vivo, in a short time, the biological potency of hepatoprotective antioxidants.

Acknowledgment. Thanks go to Miss M. Fausel and Miss S. Heidinger for skilful technical assistance.

References

1. Brehe, J.E., Burch, H.: Enzymatic assay of glutathione. Analyt. Biochem. 74, 189-197 (1976)
2. Chasseaud, L.F.: In: Glutathione: metabolism and function: GSH conjugates and mercapturic acids. Arias, I.M., Jakoby, W.B. (eds.), pp.77-114. New York: Raven Press 1976
3. Christophersen, B.O.: Reduction of linolenic acid hydroperoxide by a glutathione peroxidase. Biochim. Biophys. Acta 176, 463-470 (1969)
4. Hafeman, P.G., Hoekstra, W.G.: Protection against carbon tetrachloride-induced lipid peroxidation in the rat by dietary vitamin E, selenium and methionine as measured by ethane evolution. J. Nutr. 107, 656-665 (1977)
5. Heath, R.L., Packer, L.: Effect of light on lipid peroxidation in chloroplasts. Biochem. Biophys. Res. Commun. 19, 716-720 (1965)
6. Higashi, T., Tateishi, N., Naruse, A., Sakamoto, Y.: Decrease of glutathione and induction of γ-glutamyltransferase by dibutyryl-3',5'-cyclic AMP in rat liver. Biochem. Biophys. Res. Commun. 68, 1280-1286 (1976)
7. Jollow, D.J., Mitchell, J.R., Potter, W.Z., Davis, D.Z., Gillette, J.R., Brodie, B.B.: Acetaminophen-induced hepatic necrosis. II. Role of covalent binding in vivo. J. Pharmacol. Exp. Ther. 187, 195-202 (1973)
8. Litov, R.E., Dillard, D.J., Downey, J., Irving, D., Sagai, M., Gee, D., Tappel, A.L.: Measurement of in vivo lipid peroxidation via pentane: effect of vitamin E, selenium and polyunsaturated fat and assay for acute toxic reactions. Federation Proc. 37, 706 (1978)
9. Lowry, O.H., Rosebrough, N.J., Lewis, F., Randall, R.J.: Protein measurement with the folin phenol reagent. J. Biol. Chem. 193, 265-273 (1951)
10. Mitchell, J.R., Jollow, D.J., Potter, W.Z., Gillette, J.R., Brodie, B.B.: Acetaminophen-induced hepatic necrosis. IV. Protective role of glutathione. J. Pharmacol. Exp. Ther. 187, 211-217 (1973)
11. Recknagel, R.O.: Carbon tetrachloride hepatotoxicity. Pharmacol. Rev. 19, 145-208 (1967)
12. Riely, C.A., Cohen, G., Lieberman, M.: Ethane evolution: a new index of lipid peroxidation. Science 183, 208-210 (1974)
13. Sies, H., Gerstenecker, Ch., Menzel, H., Flohé, L.: Oxidation in the NADP system and release of GSSG from hemoglobin-free perfused rat liver during peroxidatic oxidation of glutathione by hydroperoxides. FEBS Lett. 27, 171-175 (1972)
14. Slater, T.F.: Necrogenic action of carbon tetrachloride in the rat: a speculative mechanism based on activation. Nature (London) 209, 36-40 (1966)
15. Tateishi, H., Higashi, T., Shinya, S., Naruse, A., Sakamoto, Y.: Studies on the regulation of glutathione level in rat liver. J. Biochem. 75, 93-103 (1974)

Cellular Functions in Glutathione-Depleted Hepatocytes*

J. HÖGBERG, I. ANUNDI, A. KRISTOFERSON, and A. H. STEAD

Summary

These studies were performed under the general hypothesis that glutathione (GSH) deficiency per se is not compatible with cell survival. Isolated hepatocytes were incubated with halogenated acetamides to deplete the GSH pool. The selectivity of the GSH depletion was increased by incubating at zero instead of at 37°C and after 1 h more than 50 nmol GSH was lost and less than 0.5 nmol acetamide/10^6 cells was covalently bound to cellular macromolecules. Upon continued incubation (at 37°C) in an amino acid-free medium (under these conditions the rate of GSH replenishment was absent or very low) the cells eventually lysed. A first stage of cell deterioration was detected as an inability of the cells to utilize methionine for GSH synthesis when this amino acid was added to the medium. A second stage was characterized by an accumulation of malondialdehyde in the cells, indicating an increase in the extent of lipid peroxidation. During a third stage the cells lysed. This sequence of events could be prevented by methionine when added prior to the first state or by cysteine when added prior to or during the first stage. The antioxidant α-tocopherol also prevented cell damage. The possibility that the impaired methionine utilization is a "point of no return" in a sequence which starts with GSH depletion and ends in cellular necrosis is discussed.

A conspicuous aspect of hepatic glutathione (GSH) turnover is its relationship to cell damage. Brodie and coworkers have demonstrated that bioactivated xenobiotics can deplete the hepatic GSH reservoir. During such a state of GSH deficiency drug metabolites bind to macromolecules and liver necrosis follows (1). In analogy with the concept of tumor induction and DNA alkylation, they suggested acute toxicity under such conditions to be a consequence of protein alkylation.

The analgesic drug paracematol (acetaminophen) can induce liver necrosis in many species, preceeded by a GSH deficiency (2), and the search for a rational basis for the treatment of intoxications in man has received particular attention. The plasma levels of two amino acid precursors for GSH synthesis, cysteine and methionine, are both relatively small (in the range of 0.03 mM) (3) and efforts to increase the plasma pool of sulfur amino acids by intravenous administration of methionine is a preferred treatment in not too far advanced cases of poisoning (4). However, compounds that possibly will not act as precursors for GSH synthesis have also been shown to be effective antidotes. Thus cysteamine is chosen in those cases where methionine is without effect (4). In laboratory animals known antioxidants have been found to protect against paracetamol-induced damage (5) and α-mercaptopropionylglycine protects mice without affecting either GSH synthesis or covalent binding (6).

---

*Supported by grants from Kardinska Institutet (No. 390-5) and Arbetarskyddsfonden (No. 77-295).

It is well known that GSH may act as an antioxidant and protect from lipoperoxidation damage in vitro (7) and the question was raised whether GSH deficiency per se will lead to cell damage. Suspensions of isolated hepatocytes offer possibly unique facilities to follow sequences of events leading to cell damage and a study was undertaken to characterize the effect of GSH depletion in such a system.

Rat hepatocytes were isolated by a collagenase perfusion technique and incubated in rotating flasks at 37°C (8). The medium was a Krebs-Henseleit buffer, supplemented with Hepes buffer (12.5 mM) and horse serum (5% v/v). Cellular damage was detected as an increased permeability to NADH (8). The level of acid-soluble thiols, taken as a measure of GSH, was estimated by the Saville method (9) in sedimented cells. Malondialdehyde was determined in the total incubate (7) by the thiobarbituric acid method.

In order to obtain a rapid GSH depletion, compounds that conjugate GSH without prior bioactivation have been used. Diethylmaleate is a well-known GSH-depleting agent (10), but it is poorly water soluble. With halogenated acetamides, which are readily soluble in the incubation medium, a more reproducible result was obtained. Some characteristics of cells incubated at zero degrees for 1 h in the presence of [^{14}C]-labeled iodoacetamide (0.075 mM) are given in Table 1. By this treatment a quite selective effect on acid soluble thiols was obtained. More than 50 nmol was lost during the period in which less than 0.5 nmol iodoacetamide was covalently bound to cellular macromolecules. After the first hour of incubation no free iodoacetamide was detectable in the medium, and more than 90% of the iodoacetamide taken up by the cells was recovered as the GSH conjugate (11). No signs of cell damage were observed after the first hour.

In an amino-acid free medium GSH-depleted hepatocytes eventually lysed, registered as an increased cell permeability [Fig.1; cf. (8)]. This increase in permeability was preceded by an accumulation of malondialdehyde in the cells. Several observations indicated that lipid peroxidation was the actual cause of cell lysis. Thus, malondialdehyde did not accumulate in cells lysed by detergents or freezing and thawing, while α-tocopherol or menadione prevented both malondialdehyde accumulation and cell lysis (11). Before any malondialdehyde accumulated in the cells the GSH level had to be close to zero for a certain time period (Fig.2). The incubation time required to induce malondialdehyde accumulation not only varied with the amount of GSH-depleting agent added but also, to a slight extent, with the different cell preparations.

Table 1. Effect of iodoacetamide treatment of isolated hepatocytes. The table shows some cellular characteristics before and after |^{14}C|-iodoacetamide (0.075 mM) treatment for 1 h. The temperature was kept at zero degrees to minimize iodoacetamide binding to macromolecules (11). The amount of thiols and radiolabel was determined in cells separated from the medium by centrifugation (80 *g*). The acid-precipitated cells were washed in methanol (13) prior to counting

Time (min)	Acid soluble thiols	|^{14}C|-iodoacetamide bound to acid precipitate	Permeability (% of total)
	(nmol/10^6 cells)		
0	55	0	18
60	3	0.4	17

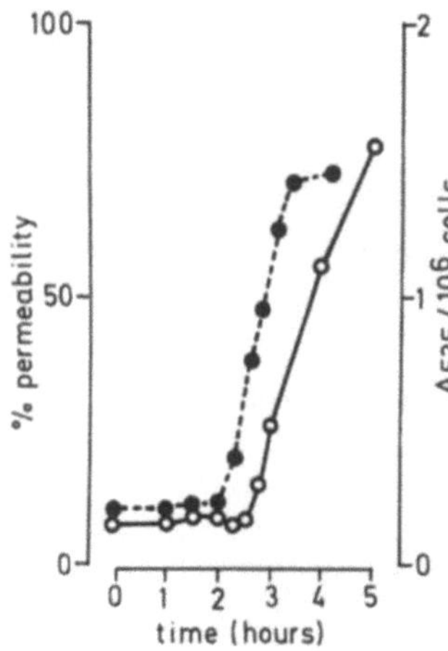

Fig. 1

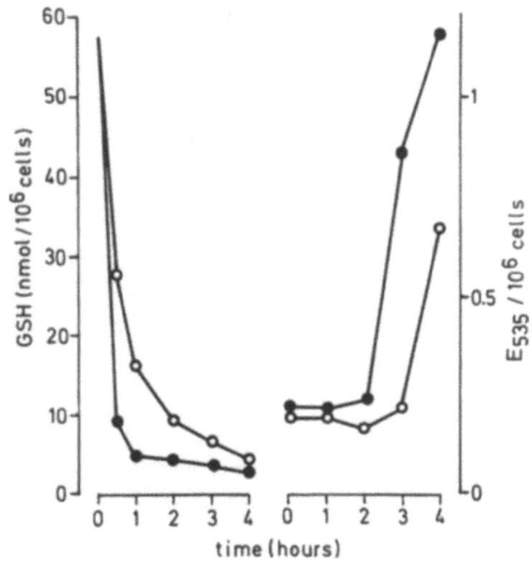

Fig. 2

Fig. 1. Effect of GSH depletion by iodoacetamide on isolated hepatocytes. Cells were incubated with iodoacetamide (0.075 mM) added to the amino acid-free medium. Malondialdehyde (●---●) and cellular permeability (o——o) was determined in aliquots of the total incubate

Fig. 2. Effect of chloroacetamide on GSH levels and malondialdehyde accumulation. Cells were incubated with two different concentrations of chloroacetamide, 0.1 mM (o——o) and 0.2 mM (●——●), and the levels of acid-soluble thiols and malondialdehyde were determined.

Addition of methionine to the medium rapidly led to an increased GSH synthesis (12) and this amino acid fully protected against iodoacetamide-induced malondialdehyde accumulation when added at zero time (Fig.3). The fact that methionine protected against lipid peroxidation after the first hour is positive evidence that GSH deficiency might be the cause of lipid peroxidation. At this time point no further iodoacetamide interactions can be expected to occur in the cells since the added amount, as mentioned above, is already metabolized. Thus, the protein binding observed during the first hour seems to be without effect on cell viability as the hepatocyte can be saved by a treatment that possibly will not affect protein alkylation.

Lipid peroxidation was not inhibited by methionine when added after two hours (Fig.3). However, further experiments showed that cysteine would "rescue" cells in situations where methionine could not (Fig.4). The lack of effect of methionine could be ascribed to a decreased methionine uptake by the hepatocytes and a decreased methionine dependent thiol production (not documented).

Methionine is taken up by a plasma membrane transport mechanism which seems to be separated from, for example, the cysteine uptake mechanism, and the methionine transformation to cysteine, once inside the cell, is a function of the "transsulfuration pathway" (12). A selective damage somewhere in either the membrane transport or the transsulfuration system thus precedes the signs of lipid peroxidation.

We do not know the explanation for the impaired methionine utilization and can only speculate about the in vivo significance of this finding. Comparative studies on cysteine and methionine uptake characteristics

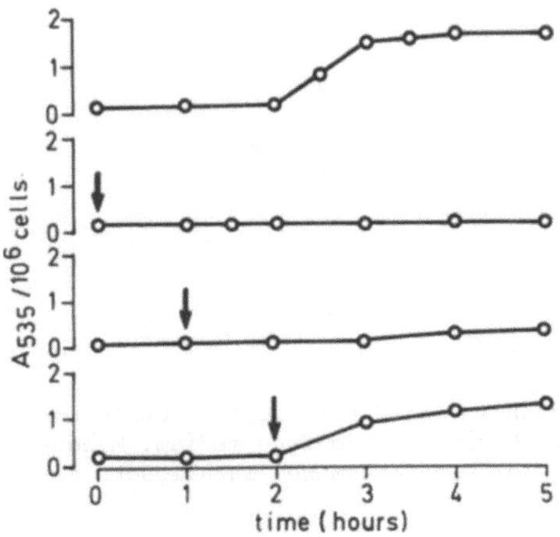

Fig. 3

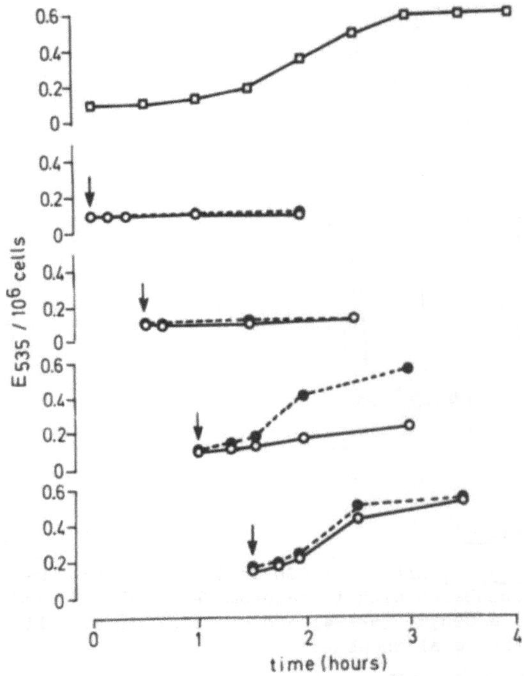

Fig. 4

Fig. 3. Effect of methionine addition to chloroacetamide treated hepatocytes on malondialdehyde formation. Cells were incubated in four different flasks, all containing chloroacetamide (0.2 mM). Methionine (0.2 mM) was added at time points indicated by *arrows* to three of the flasks

Fig. 4. Effect of cysteine and methionine on iodoacetamide induced lipid peroxidation. Cells were incubated in an amino acid-free stock medium (180 ml), containing iodoacetamide (0.1 mM). Two aliquots (15 ml each) were withdrawn in pairs at different times (indicated in the figure) and to the first cysteine (o-o) was added and to the second methionine (●---●) was added. The incubation was then continued and the level of malondialdehyde was measured in both the stock solution (□——□) and the aliquots. The amino acid concentration was 1 mM

indicate that cysteine in plasma has to compete with many other amino acids to be taken into hepatocytes, while methionine uptake was more or less unaffected by other amino acids (12). Thus, transport mechanisms may favor a methionine utilization such that the plasma pool of this amino acid is more important for hepatic cysteine turnover than cysteine itself (12). A shut-down in methionine utilization may thus be of crucial importance in a situation where the rate of GSH synthesis can be expected to determine whether the cell will survive or not.

References

1. Gillette, J.R., Mitchell, J.R., Brodie, B.B.: Biochemical mechanisms of drug toxicity. Ann. Rev. Pharmacol. 14, 271-289 (1975)
2. Davis, D.C., Potter, W.Z., Jollow, D.J., Mitchell, J.R.: Species differences in hepatic glutathione depletion, covalent binding and hepatic necrosis after acetaminophen. Life Sciences 14, 2099-2109 (1974)

3. Brigham, M.P., Stein, W.H., Moore, S.: The concentrations of cysteine and cystine in human blood plasma. J. Clin. Invest. 39, 1633-1638 (1960)
4. Prescott, L.F., Park, J., Proudfoot, A.T.: Cysteamine, L-methionine and D-penicillamine in paracetamol poisoning. J. Int. Med. Res. 4, Suppl.4, 112-117 (1976)
5. Kelleher, J., Keaney, N.P., Walker, B.E., Losowsky, M.S.: Modification of paracetamol hepatotoxicity by anti-oxidants. J. Int. Med. Res. 4, Suppl.4, 139-144 (1976)
6. Labadarios, D., Davis, M., Portman, B., Williams, R.: Paracetamol-induced hepatic necrosis in mouse, relationship between covalent binding, hepatic glutathione depletion and the protective effect of α-mercaptopropionylglycine. Biochem. Pharm. 26, 31-35 (1977)
7. Högberg, J., Orrenius, S., Larson, R.E.: Lipid peroxidation in isolated hepatocytes. Eur. J. Biochem. 50, 595-602 (1975)
8. Högberg, J., Kristoferson, A.: A correlation between glutathione levels and cellular damage in isolated hepatocytes. Eur. J. Biochem. 74, 77-82 (1977)
9. Saville, B.: A scheme for the colorimetric determination of microgram amounts of thiols. The Analyst 83, 670-672 (1958)
10. Chasseaud, L.F.: Glutathione S-transferases. In: Glutathione. Flohé, L., Benhör, H.C., Sies, H., Waller, H.D., Wendel, A. (eds.), pp.90-109. Stuttgart: Georg Thieme 1974
11. Anundi, I., Högberg, J., Stead, A.H.: Glutathione depletion in isolated hepatocytes; its relation to lipid peroxidation and cell damage. (Submitted for publication)
12. Högberg, J., Kristoferson, A.: Methionine and cysteine transport in hepatocyte suspensions and its relation to hepatic glutathione synthesis. (Manuscript in preparation)
13. Reid, W.D., Krishna, G., Gillette, J.R., Brodie, B.B.: Biochemical mechanism of hepatic necrosis induced by aromatic hydrocarbons. Pharmacology 10, 193-214 (1973)

Glutathione and Glutathione Conjugate Metabolism in Isolated Liver and Kidney Cells

D. P. JONES, A. H. STEAD, P. MOLDÉUS, and S. ORRENIUS

Summary

The metabolism of extracellular glutathione and paracetamol-S-glutathione conjugate has been studied in isolated liver and kidney cells. With kidney cells, but not with liver cells, added GSH is rapidly oxidized. The oxidation is O_2-dependent and yields GSSG, which is the substrate for subsequent degradative reactions. GSSG is only slowly metabolized by liver cells, but this activity is increased following pretreatment with phenobarbital. By comparison, the rate in untreated kidney cells is considerably higher. Analysis of the metabolites by high-pressure liquid chromatography suggests that the preferred route of GSSG breakdown includes (1) removal of a γ-glutamyl residue, (2) removal of the glycyl residue from the cysteinylglycine moiety, (3) removal of the second γ-glutamyl residue, and (4) removal of the second glycyl residue to yield cystine.

Upon incubation of liver cells with paracetamol, the GSH consumed is nearly quantitatively recovered as the glutathione derivative. Addition of isolated kidney cells to the liver incubate results in conversion of this metabolite to the cysteine and N-acetylcysteine derivatives. This conversion is inhibited by addition of GSSG. Addition of alanylglycine, a substrate for the peptidase which hydrolyses cysteinylglycine, results in accumulation of another intermediate, paracetamol-S-cysteinylglycine. Thus, the reaction sequence in metabolism of glutathione conjugates appears to be the same as that for glutathione breakdown.

Metabolic activities related to glutathione function are partitioned between various tissues. Thus, we know that the liver releases GSSG under various conditions (1,9,14-16), and apparently releases GSH continually (1,16), yet does not appear to remove these substances from the blood (4). On the other hand, the kidney is known actively to remove GSH from the perfusate during perfusion, and contains enzymes such as γ-glutamyltransferase (EC 2.3.2.2) and cysteinylglycinase (EC 3.4.13.6) which appear to have an important role in glutathione turnover (3,10).

Our interest in glutathione metabolism arose initially from our studies of drug metabolism in the liver. Using isolated liver cells Moldéus (12) found that glutathione is essentially quantitatively recovered as paracetamol-glutathione after 5 h incubation with paracetamol. Only very small amounts of paracetamol-cysteine and no paracetamol-N-acetylcysteine were detected despite the fact that urinary products following paracetamol administration in vivo include the latter two derivatives only (12). Conversion of glutathione conjugates of drugs to cysteine derivatives is known to occur in kidney (2,19) in reactions which also appear to involve γ-glutamyltransferase (11,18) and cysteinylglycinase (6).

We recently developed an isolated kidney cell preparation for drug metabolism and toxicity studies (8). This system consists of a mixture of isolated intact kidney cells derived from rat kidney by perfusion with a solution containing collagenase. The final cell preparation is comparable to freshly isolated liver cells in viability parameters (8) and is active in metabolism of paracetamol. However, the rates of metabolite formation are considerably less than those found with liver cells. The profile of paracetamol metabolites shows no glutathione conjugate, but both cysteine and N-acetylcysteine conjugates. Presumably, paracetamol-glutathione is formed in the kidney cells in a manner analogous to that which occurs in the liver cells; however, in the kidney cells, subsequent reactions can occur which convert the glutathione derivative to the other derivatives.

To test this hypothesis, we incubated isolated liver cells from phenobarbital pretreated rats with paracetamol to form paracetamol-glutathione (13). The isolated liver cells were then removed from the suspension by centrifugation and isolated kidney cells were added. The results (Fig.1) show a rapid conversion of the paracetamol-glutathione to paracetamol-cysteine and the subsequent acetylation of this to form paracetamol-N-acetylcysteine (13). The sequence of events thus seems to be as described in Figure 2, where paracetamol is oxidized to a reactive electrophile which combines with GSH in the liver. The conjugate is then a substrate for the kidney enzymes, γ-glutamyltransferase and cysteinylglycinase, and is converted to paracetamol-cysteine. This is

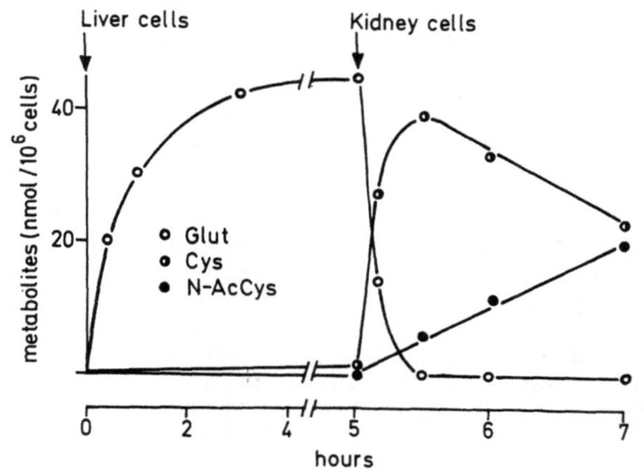

Fig. 1. Kidney cell metabolism of paracetamol-glutathione. Paracetamol-glutathione was formed by incubation of isolated liver cells with paracetamol. Liver cells were then removed by centrifugation and kidney cells were added as indicated. [Data from (13)]

Liver

paracetamol —1→ reactive electrophile —2→ paracetamol glutathione

Kidney

paracetamol glutathione —3→ paracetamol cysteinylglycine —4→ paracetamol cysteine —5→ paracetamol N-acetylcysteine

Fig. 2. Sequence of reactions in paracetamol metabolism. Reactions occurring primarily in the liver: 1, cytochrome P-450-dependent oxidation to a reactive electrophilic compound; 2, conjugation either enzymatically or nonenzymatically with GSH. Reactions occurring primarily in the kidney: 3, removal of γ-glutamyl residue by γ-glutamyltransferase; 4, removal of glycyl residue by cysteinylglycinase; 5, N-acetylation by N-acetyltransferase

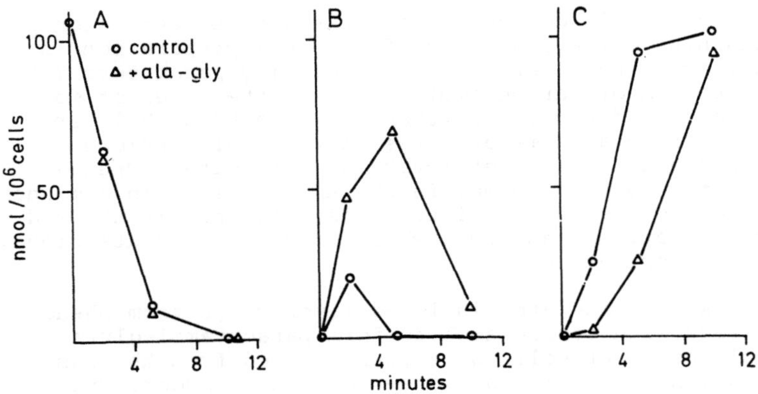

Fig. 3A-C. Metabolism of paracetamol-glutathione in isolated kidney cells in the presence of alanylglycine. (A) Breakdown of paracetamol-glutathione. (B) Accumulation of paracetamol-cysteinylglycine. (C) Accumulation of paracetamol-cysteine. [Data from (7)]

subsequently N-acetylated to form the mercapturic acid derivative. By use of a dipeptide, alanylglycine, as a competitive inhibitor of the dipeptidase which hydrolyses the glycine residue, we have been able to force accumulation of the cysteinylglycine derivative of paracetamol, which we can separate and quantitate by high-pressure liquid chromatography (Fig.3).

To gain a better understanding of the relationship of the metabolism of glutathione conjugates with the γ-glutamyltransferase reaction, we studied glutathione metabolism by kidney cells (7).

Our first experiment was to examine GSH loss with isolated kidney cells. Since the methods which measure total thiol would not detect a simple conversion of γ-glutamylcysteinylglycine to cysteinylglycine or cysteine, we first used the o-phthalaldehyde method of Hissin and Hilf (5). The results were as anticipated; there was a rapid loss of GSH (Fig.4A). Subsequent measurement of GSH as total thiol gave an identical loss, which suggested that the free sulfhydryl was being lost at the same time. We then measured GSSG, also by the Hissin and Hilf method, and found that GSSG increase initially was nearly stoichiometric with GSH loss (Fig.4B). After the initial 10-15 min, during which time the GSH was totally converted to GSSG, the GSSG concentration decreased linearly.

We performed several controls for this GSH oxidation including incubations without cells, with liver cells, and with kidney cells which had been heat-denatured. All these experiments showed only a very slow rate of oxidation - less than 5% of the rate using intact kidney cells (7). The reaction was found to be dependent upon O_2 and inhibited by KCN, but to be unchanged by the respiratory inhibitor Antimycin A (Fig.5A). The very high rate of GSH oxidation under our assay conditions was due to an unusual substrate dependence (Fig.5B) where the rate was greater than first order with respect to GSH concentration. In light of the very high rate of the reaction and the recent observation that GSH is released from liver at a considerable rate (15 nmol/min/g) (16), it appears that this oxidative activity may have a physiological role in GSH metabolism. The GSSG formed may be the true substrate for degradation reactions.

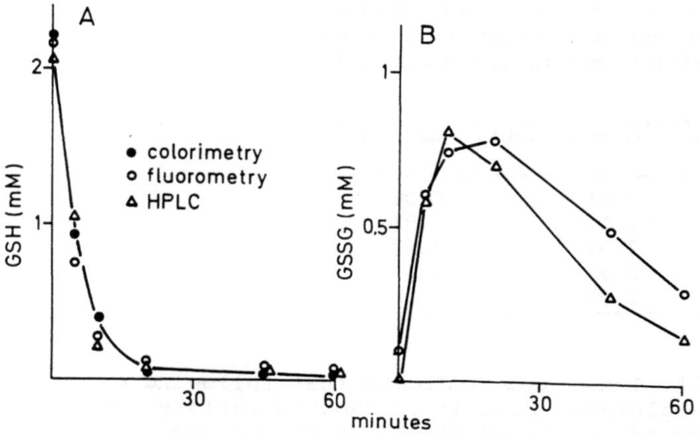

Fig. 4A and B. Oxidation of exogenous GSH by isolated kidney cells. Cells were incubated (10^6 cells/ml) in Krebs-Henseleit buffer. Incubations without cells, with boiled cells or with control liver cells showed only very slow oxidation. [Data from (7)]

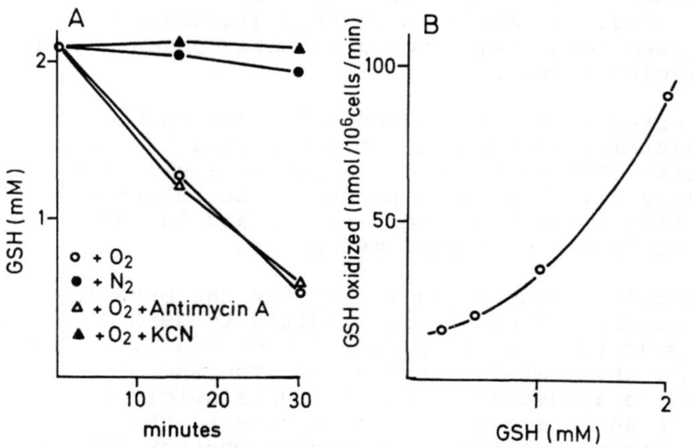

Fig. 5A and B. Dependence of GSH oxidation on O_2 (A) and GSH (B). Conditions as in Figure 4 with additions as indicated. [Data from (7)]

To study this we used a high-pressure liquid chromatographic method, initially developed by Stead and Högberg (17) for quantitation of GSH and GSSG (see Fig.4), to separate any metabolites formed during GSH or GSSG breakdown. Starting with either GSH or GSSG as substrate, we found eight principal peaks (7), two of which we knew to be GSH and GSSG. The only difference between the time course of formation of metabolites starting with GSH as opposed to GSSG was that the initial metabolite formation was delayed with GSH as substrate, presumably due to the time required for oxidation.

We collected the individual peaks and analyzed them after acid hydrolysis for amino acid content. The peaks contained amino acids as shown in Table 1. Preparation of mixed disulfides from cysteine, cysteinylglycine and glutathione confirmed the peptides given in the sequence in Figure 6. The time course of formation of the metabolites is shown

Table 1. Amino acid composition of peaks collected following high-pressure liquid chromatographic separation of the supernatants from kidney cells which had been incubated with GSH. [Data from (7)]

Peak	Retention time (s)	Glutamate	Cysteic acid	Glycine
II	950	0.00	2.00	0.06
III	1020	0.18	2.00	0.93
IV	1060	1.09	2.00	1.17
V	1270	1.06	2.00	2.31
VI	1326	0.00	2.00	2.12

schematically in Figure 7A. These data indicate that following oxidation of GSH to GSSG, a γ-glutamyl residue is removed rapidly and a steady-state concentration of the mixed disulfide of GSH and cysteinylglycine is reached. Subsequently, either the glycine from the cysteinylglycine moiety or the other γ-glutamyl residue is removed, but the glycine hydrolysis appears to be the preferred route. The subsequent step from either the mixed disulfide of GSH and cysteine (GSH-Cys) or cystinyl-bis-glycine (Cysgly-Cysgly) yields the mixed disulfide of cysteine and cysteinylglycine (Cys-Cysgly; see Fig.6). The final glycyl residue is hydrolyzed to yield glycine and cystine, which accumulate essentially quantitatively (Fig.7B). Analysis of free glutamate and glycine supports this proposed sequence, since glutamate concentration rises more rapidly than glycine (Fig.7B).

GSSG metabolism is not detected in control liver cells presumably because of deficiency in γ-glutamyltransferase activity. In cells from phenobarbital pretreated rats GSSG is metabolized but at a slow rate compared to kidney cells (Fig.8). This corresponds to the relative γ-glutamyltransferase activities under these conditions, and further suggests that this enzyme is involved in GSSG metabolism.

These results suggest a reaction sequence in kidney for the breakdown of GSSG, and also of GSH, assuming that GSH is oxidized to GSSG in vivo. The same reactions also appear to occur in conversion of the glutathione derivative of paracetamol to the cysteine derivative. The use of a competitive inhibitor resulted in accumulation of the cysteinylglycine derivative of paracetamol as an intermediate. To examine further the relationship of these two systems, we therefore tested whether GSSG is also a competitive inhibitor of the conversion of paracetamol-glutathione to other derivatives. The results (Fig.9) showed that GSSG is a rather good inhibitor of this process. These two inhibitor experiments suggest that the two reaction sequences both involve γ-glutamyltransferase and cysteinylglycinase.

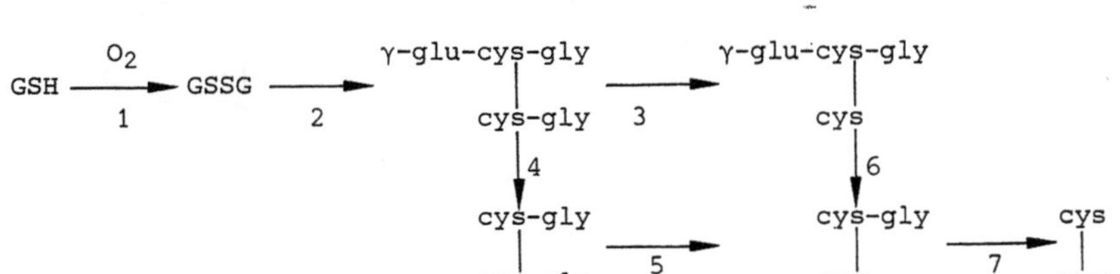

Fig. 6. Sequence of reactions in GSH metabolism by isolated kidney cells. 1, O_2-dependent oxidation; 2, 4, and 6, removal of γ-glutamyl residue by γ-glutamyl transferase; 3, 5, and 7, removal of glycyl residue by cysteinylglycinase

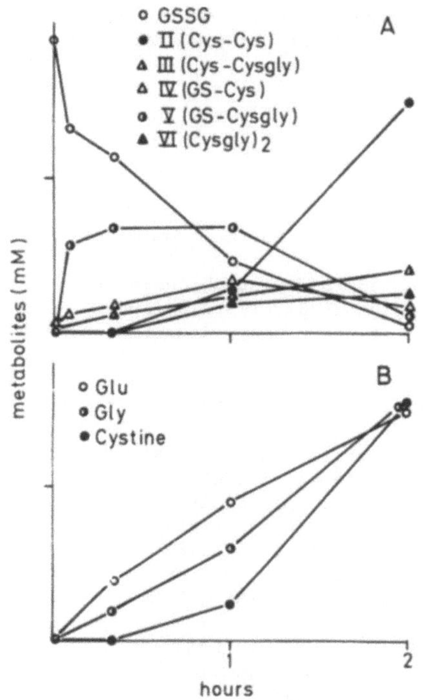

Fig. 7A and B. Time sequence of accumulation of metabolites from GSSG. (A) Concentration estimates based upon absorbance at 220 nm given in (7). (B) Amino acid accumulation

In conclusion, the reaction sequences involved in metabolism of glutathione and a glutathione conjugate by isolated kidney cells have been studied. Intermediates have been isolated which suggest sequential reactions involving γ-glutamyltransferase and cysteinylglycinase for both of these processes, and competition experiments suggest that the same enzymes are utilized for both processes. Rapid oxidation of GSH by kidney cells suggest that GSSG is the true substrate for the degradative reactions, even though the isolated enzymes have higher affinity for the thiol form (10).

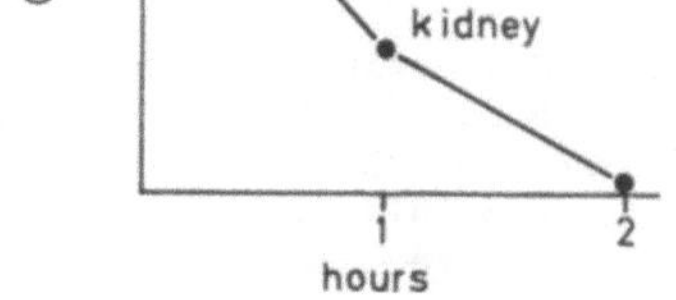

Fig. 8

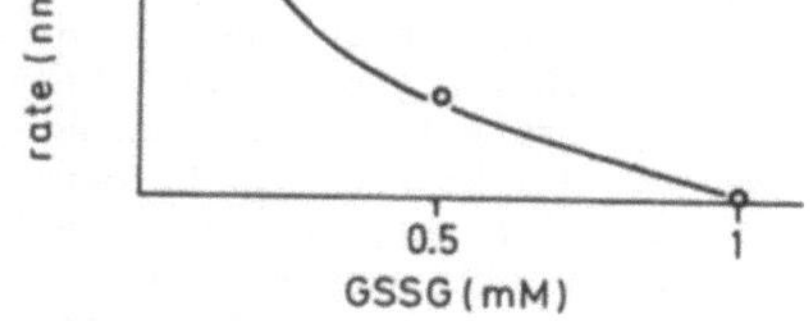

Fig. 9

Fig. 8. Loss of exogenous GSSG in incubations of hepatocytes and kidney cells. Cells were incubated at 10^6/ml in Krebs-Henseleit buffer

Fig. 9. GSSG inhibition of rate of paracetamol glutathione breakdown. [Data from (7)]

Acknowledgments. These studies were supported by a grant from the Swedish Medical Research Council (proj. O3X-2471). One of the authors (A.H.S.) was the recipient of a Council of Europe Fellowship.

References

1. Bartoli, G.M., Sies, H.: Reduced and oxidized glutathione efflux from liver. FEBS Lett. 86, 89-91 (1978)
2. Boyland, E., Chasseaud, L.F.: The role of glutathione and glutathione-S-transferases in mercapturic acid biosynthesis. Adv. Enzymol. 32, 173-219 (1969)
3. Elce, J.S., Broxmeyer, B.: γ-Glutamyltransferase of rat kidney. Simultaneous assay of the hydrolysis and transfer reactions with |Glutamate-^{14}C| glutathione. Biochem. J. 153, 223-232 (1976)
4. Hahn, R., Wendel, A., Flohé, L.: The fate of extracellular glutathione in the rat. Biochim. Biophys. Acta 539, 324-337 (1978)
5. Hissin, P.J., Hilf, R.: A fluorometric method for determination of oxidized and reduced glutathione in tissues. Anal. Biochem. 74, 214-226 (1976)
6. Hughey, R.P., Rankin, B.B., Elce, J.S., Curthoys, N.P.: Specificity of a particulate rat renal peptidase and its localization along with other enzymes of mercapturic acid synthesis. Arch. Biochem. Biophys. 186, 211-217 (1978)
7. Jones, D.P., Moldéus, P., Stead, A.H., Ormstad, K., Jörnvall, H., Orrenius, S.: Metabolism of glutathione and a glutathione conjugate by isolated kidney cells. (Submitted for publication, 1978)
8. Jones, D.P., Sundby, G.-B., Ormstad, K., Orrenius, S.: Use of isolated kidney cells for study of drug metabolism. Biochem. Pharmacol., in press
9. Jones, D.P., Thor, H., Andersson, B., Orrenius, S.: Detoxification reactions in isolated hepatocytes. Role of glutathione perocidase, catalase and formaldehyde dehydrogenase in reactions relating to N-demethylation by the cytochrome P-450 system. J. Biol. Chem. 253, 6031-6037 (1978)
10. Meister, A.: Biochemistry of glutathione. In: Metabolism of sulfur compounds. Greenberg, D.M. (ed.), pp.101-188. New York: Academic Press 1975
11. Meister, A., Tate, S.S.: Glutathione and related γ-glutamyl compounds: biosynthesis and utilization. Ann. Rev. Biochem. 45, 559-604 (1976)
12. Moldéus, P.: Paracetamol metabolism and toxicity in isolated hepatocytes from rat and mouse. Biochem. Pharmacol., in press (1978)
13. Moldéus, P., Jones, D.P., Ormstad, K., Orrenius, S.: Formation and metabolism of a glutathione-S-conjugate in isolated rat liver and kidney cells. Biochem. Biophys. Res. Commun. 83, 195-200 (1978)
14. Oshino, N., Chance, B.: Properties of glutathione release observed during reduction of organic hydroperoxide, demethylation of aminopyrine and oxidation of some substances in perfused rat liver, and their implications for the physiological function of catalase. Biochem. J. 162, 509-522 (1977)
15. Sies, H., Gerstenecker, C., Menzel, H., Flohé, L.: Oxidation in the NADP system and release of GSSG from hemoglobin-free perfused rat liver during peroxidatic oxidation of glutathione by hydroperoxides. FEBS Lett. 27, 171-175 (1972)
16. Sies, H., Bartoli, G.M., Burk, R.F., Waydhas, C.: Glutathione efflux from perfused rat liver after phenobarbital-treatment, during drug oxidations, and in selenium deficiency. Eur. J. Biochem. 89, 113-118 (1978)
17. Stead, A.H., Högberg, J.: Determination of oxidized and reduced glutathione by high-pressure liquid chromatography. (Submitted for publication, 1978)
18. Wendel, A., Heinle, H., Silbernagl, S.: The degradation of glutathione derivatives in the rat kidney. Hoppe-Seyler's Z. Physiol. Chem. 358, 1413-1414 (1977)
19. Wood, J.L.: Biochemistry of mercapturic acid formation. In: Metabolic conjugation and metabolic hydrolysis. Fishman, W.H. (ed.), pp.261-299. London and New York: Academic Press 1970

Glutathione-Requiring Enzymes in the Metabolism of Prostaglandin Endoperoxides

E. CHRIST-HAZELHOF and D. H. NUGTEREN

Summary

Four enzymes requiring glutathione for their action are known in prostaglandin endoperoxide metabolism. Two of them are specifically required for the biosynthesis of certain prostaglandins:

1. Prostaglandin endoperoxide E-isomerase present, e.g., in sheep vesicular glands and kidney medulla, which converts the endoperoxide into prostaglandin E.
2. Prostaglandin endoperoxide D-isomerase isolated from rat spleen, which gives a high yield of prostaglandin D.

In addition, two less specific enzymes affect the prostaglandin endoperoxides:

3. The glutathione S-transferases, converting the endoperoxides into mixtures of prostaglandin F, E, and D, the composition of the mixture depending on the transferase involved: e.g., the sheep lung enzyme mainly gives a mixture of D and F.
4. Glutathione peroxidase which can convert 15-hydroperoxy prostaglandins into 15-hydroxy prostaglandins. However, this enzyme does not act on the endoperoxide group of prostaglandin endoperoxides.

A. Introduction

In our laboratory, research is done on the metabolism of essential fatty acids. Essential fatty acids are specific unsaturated fatty acids which cannot be synthesized by mammals, but are needed for optimal growth and normal function of the animal as was found by Burr and Burr in the thirties. The most important essential fatty acid is linoleic acid. Linoleic acid is a necessary precursor in mammals for the biosynthesis of dihomo-γ-linolenic acid (all-cis 8,11,14-eicosatrienoic acid) and arachidonic acid. These two fatty acids are necessary precursors in the biosynthesis of prostaglandins, hormone-like compounds which in trace amounts have profound effects on a great number of physiological activities.

Much work has already been done on the "classical" prostaglandins of the E and F series. In 1964, reduced glutathione was recognized almost immediately as a necessary cofactor in prostaglandin biosynthesis. In this paper we will discuss where and how glutathione is involved in these conversions, but first a very brief summary of the most important biosynthetic reactions presently known in this field will be given.

The key intermediate in prostaglandin biosynthesis is the prostaglandin endoperoxide PGH_2. This intermediate was isolated for the first time in 1973 (7,11) and is synthesized by the enzyme prostaglandin endoperoxide synthetase (E.C. 1.14.99.1) from arachidonic acid. Since this enzyme is membrane-bound, it is difficult to purify. Some years ago

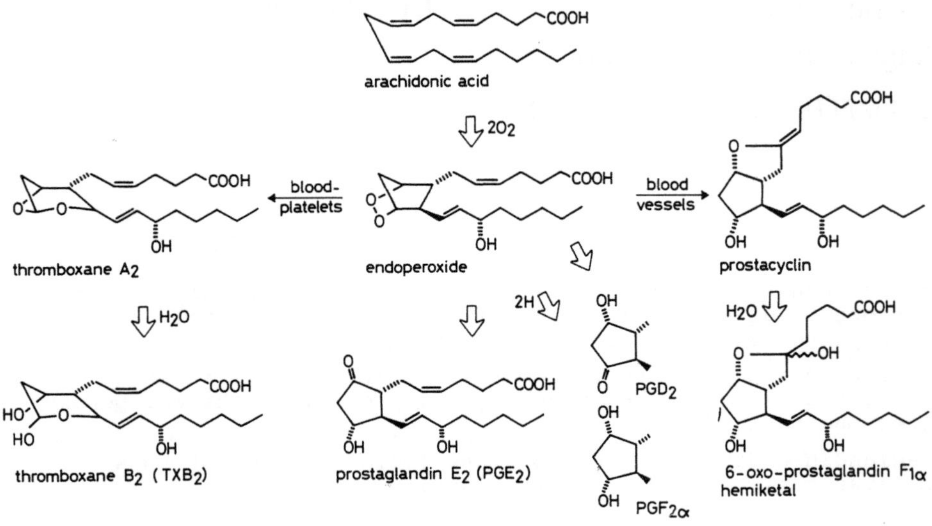

Fig. 1. Transformations of prostaglandin endoperoxide (PGH_2)

its purification and some of its properties were reported by two groups by Miyamoto et al. (10) and in our laboratory by Van der Ouderaa et al. (14). The prostaglandin endoperoxide is not stable in aqueous medium; it decomposes spontaneously into the prostaglandins E, D, and F (see Fig.1).

Under physiological conditions the prostaglandin endoperoxide is rapidly converted by a number of enzymes. Besides the formation of the "classical" prostaglandins E, F, and D, the formation of two other compounds is of interest (see Fig.1):

1. Thromboxane A_2 which is formed by a membrane-bound enzyme in blood-platelets. On hydrolysis of the oxetane-ring, thromboxane B_2 is formed (8).
2. Prostacyclin, which was recently discovered by Moncada et al. and is formed by the walls of blood vessels. Also this compound is unstable, the cyclic enol-ether quickly hydrolyzes and 6-oxo-$PGF_{1\alpha}$ is formed (9).

These two compounds, thromboxane A_2 and prostacyclin, have very interesting antagonistic properties in blood platelet aggregation. Thromboxane A_2 is a strongly pro-aggregating substance, whereas prostacyclin is a strong inhibitor of aggregation. Some of the "classical" prostaglandins E and D also have pro- and anti-aggregating properties. Thus it may be of vital importance to know which of these pro- or anti-aggregating substances are formed preferentially. To obtain a better understanding of these factors influencing these conversions we have tried to isolate and characterize some of the enzymes involved.

B. Prostaglandin Endoperoxide E-Isomerase

Prostaglandin E, one of the earliest prostaglandins known is formed from the endoperoxide by a membrane-bound enzyme: the prostaglandin endoperoxide E-isomerase (E.C. 5.3.99.3). This enzyme converts the endoperoxide into prostaglandin E almost without formation of other products such as prostaglandins F and D. Sheep vesicular glands contain large amounts of this enzyme but it has also been found in other organs, e.g., in kidney medulla, bovine and pig lung. Ogino et al. (12) studied this rather unstable enzyme. Various thiol compounds including glutathione protected the enzyme from inactivation. However, there were indications that glutathione was specifically involved as a coenzyme in the isomerisation reaction, although it was not oxidized in a stoichiometric quantity.

C. Prostaglandin Endoperoxide D-Isomerase

Another prostaglandin, namely prostaglandin D, is formed from the endoperoxide in many organs of the rat, and also in other animal species, by an enzyme called prostaglandin endoperoxide D-isomerase (E.C.5.3. 99.2). We succeeded in purifying this cytoplasmic enzyme from rat spleen until homogeneous (4). This enzyme appeared to be as specific as the E-isomerase just mentioned. It exclusively catalyzes the formation of prostaglandin D from the endoperoxide, thus almost without formation of any other products as prostaglandins E and F. Glutathione seemed again a necessary cofactor. The enzyme was also rather labile, even in the presence of 5 mM 2-mercapto-ethanol, forcing us to develop a rapid purification scheme, which could be finished in 5 days. All steps were done in the presence of 5 mM 2-mercapto-ethanol and in the cold at 0°C-4°C.

As stated above, the prostaglandin endoperoxide is an unstable compound that decomposes spontaneously in aqueous medium into a mixture of prostaglandins E, D, and F. Enzyme activity is therefore given as an increase in the quantity of D formed, either expressed as a D/E ratio or as the percentage D of total products formed. Enzyme activity was determined by incubation of [1-^{14}C] prostaglandin endoperoxide followed by quantitative determination of the products formed by liquid scintillation counting after separation on TLC.

After centrifugation of the spleen homogenate, the supernatant was subjected to DEAE (anion exchange) chromatography in a Tris buffer using a salt gradient. This was followed by a preparative isoelectric-focusing step, using a Sephadex G15 column and a pH gradient between pH 4 and 6. The iso-electric point appeared to be 5.2. The next step was affinity chromatography, and this was done analogous to a successful step in the purification of glutathione S-transferases as carried out by Simons and Van der Jagt (13). Since our enzyme needed glutathione for activity, we hoped that it would also be bound on a glutathione affinity column. Such a column was prepared by coupling glutathione with epoxy-activated Sepharose 4B. As is shown in Figure 2 this method proved extremely efficient: in one step a 200-fold purification was achieved. Practically all other proteins were directly eluted whereas the isomerase activity was completely bound to the column. It could only be eluted with a high pH buffer containing 5 mM glutathione. The purification was completed with gel filtration on Sephadex G 100. Sodium dodecyl sulphate slab gel electrophoresis proved that the enzyme was pure after the affinity-chromatography step. Its mobility compared with reference proteins revealed a molecular weight of 26,000. On gel filtration on Sephadex G 100 a molecular weight of 34,000 was found. Hence the enzyme apparently consists of one polypeptide chain of about 30,000 daltons.

With the purified enzyme it was found that no other thiol compounds or reducing agents could replace glutathione in the reaction. Figure 3

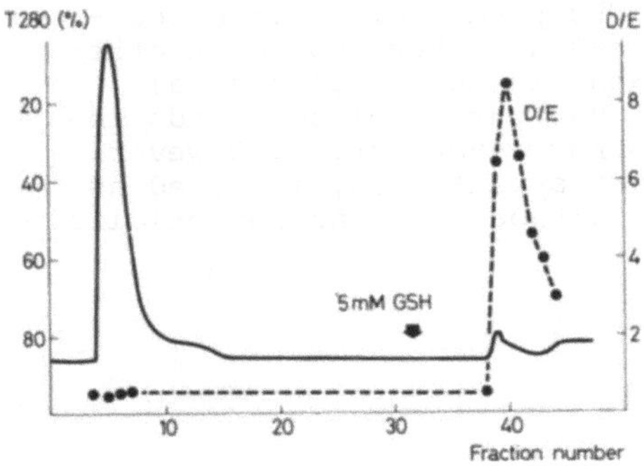

Fig. 2

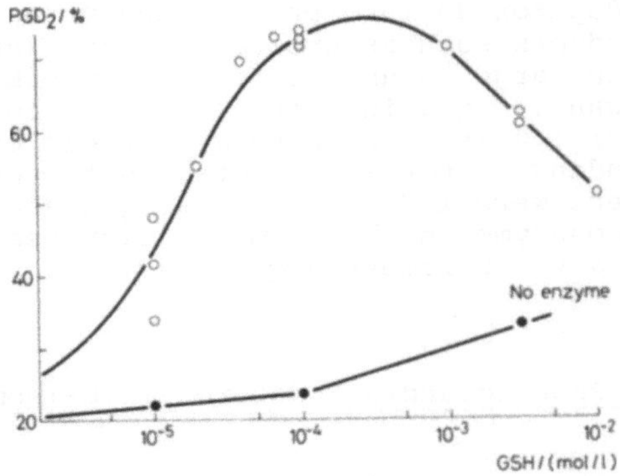

Fig. 3

Fig. 2. Glutathione affinity chromatography in the purification of prostaglandin endoperoxide D-isomerase

Fig. 3. Effect of the concentration of glutathione on the formation of prostaglandin D_2 with (-o-o-o-) and without (-●-●-●-) rat spleen D-isomerase

shows that the optimal glutathione concentration lies between 10^{-3} and 10^{-4} M. The highest conversion was obtained between pH 7 and 8 which is very similar to that found by Ogino et al. (12) for bovine prostaglandin endoperoxide E-isomerase. We also tested a number of compounds which might inhibit the isomerase either by interfering with the substrate-binding site or with the glutathione-binding site. As could be expected, p-hydroxy-mercuribenzoate did inhibit the reaction completely when equimolar concentrations of glutathione and p-hydroxy-mercuribenzoate were used. Oxidized glutathione, S-methyl-glutathione and ophthalmic acid showed weak inhibition. A number of compounds similar to the substrate were tested. These compounds, such as fatty acids, hydroxy-acids, or prostaglandin-like compounds only showed weak inhibition or no inhibition at all (4).

Because of their characteristics as outlined above, both E-isomerase and D-isomerase may be considered as quite specific; their sole function seems to be to convert the endoperoxide into prostaglandin E or D without any other products being formed.

D. Glutathione S-Transferases

When we tried to isolate the prostaglandin endoperoxide D-isomerase found in rat spleen from more practical sources as e.g., organs of slaughterhouse animals, it became clear that also glutathione-requiring enzymes with a much wider substrate specificity may be involved in prostaglandin endoperoxide metabolism. After some tests, we thought sheep lung was a good source of endoperoxide D-isomerase and possibly also of endoperoxide F-reductase, since on incubation with sheep lung homogenates quite large amounts of prostaglandin F were found besides prostaglandin D. The formation of prostaglandin F was of special interest since so far no other enzyme which catalyzes the reduction of the endoperoxide had been described. Both enzymes were present in the 100,000 g supernatant, both were precipitated with ammonium-sulphate and both were eluted simultaneously on DEAE cellulose. However, on CM-cellulose chromatography two distinct peaks appeared, but on incubation

Table 1. Incubations of [1-^{14}C]-prostaglandin endoperoxide with pure enzymes

Enzyme	Products (% radioactivity) PGD$_2$	PGE$_2$	PGF$_{2\alpha}$
None	27	65	4
Rat spleen endoperoxide D-isomerase	90	8	1
Sheep lung glutathione S-transferase	38	11	47

with prostaglandin endoperoxide both peaks again showed D-isomerase and F-reductase activity. These purified iso enzymes - about 90% pure as judged by SDS-gel electrophoresis - required glutathione for activity.

When we had come that far, a paper of Cagen et al. (1) appeared describing the formation of an adduct of glutathione and prostaglandin A_1, catalyzed by glutathione S-transferase (E.C. 2.5.1.18). These transferases add glutathione to electrophilic sites of organic compounds. Habig et al. (6) had described four glutathione S-transferases from rat liver. Our two proteins SL_1 and SL_2 both gave a glutathione adduct on incubation with prostaglandin A_1, and they also reacted with 1-chloro-2,4-dinitrobenzene and other typical glutathione S-transferase substrates. From that moment we were convinced we had isolated two new glutathione S-transferases from sheep lung. The molecular weight was 45,000, the same as was found by Habig et al. for their transferases. Like the liver transferases the sheep lung enzymes consist of two subunits with equal molecular weight. The amino acid composition of SL_2 is very similar to rat liver transferase B as determined by Habig et al. We have isolated the four rat liver transferases mentioned above and found that they also catalyze the decomposition of prostaglandin endoperoxide. All transferases gave increased F-formation, but only sheep lung transferases, in addition, catalyzed the D-formation. In Table 1 the product formation on incubation of prostaglandin endoperoxide with rat spleen prostaglandin endoperoxide D-isomerase was compared with that of sheep lung glutathione S-transferase. For further details of this work, we refer to an earlier publication (3).

The reaction product of glutathione S-transferases normally is an adduct of substrate + glutathione. The same applies when prostaglandin A is the substrate (2). If there would also be an adduct during the conversion of prostaglandin H_2 catalyzed by glutathione S-transferases, we might expect that this adduct could not be extracted into ether from an acidified aqueous solution. However, we found that the radioactivity, after acidification, was always completely recovered in the ether phase even after very short incubation times and rapid extraction at low temperature. Hence, no indications regarding adduct formation could be derived from this experiment.

E. Glutathione Peroxidase

Finally we want to discuss briefly still another glutathione-requiring enzyme in prostaglandin endoperoxide metabolism, namely glutathione peroxidase. Glutathione peroxidase catalyzes the reduction of hydrogenperoxide and organic hydroperoxides, with glutathione as hydrogen donor.

We found that also 15-hydroperoxy-prostaglandin E is rapidly reduced by this enzyme (15). Although it is generally assumed that dialkyl peroxides are not attacked by this enzyme, it could be that strained endoperoxide ring of prostaglandin H_2 behaved differently. However, on incubation with glutathione peroxidase purified according to Flohé et al. (5), we found no increased formation of prostaglandin $F_{2\alpha}$ (3).

Glutathione peroxidase is apparently not necessary in prostaglandin biosynthesis, because recent findings in our laboratory (14) and in a Japanese laboratory (10) have shown that pure, homogeneous prostaglandin endoperoxide synthetase has cyclooxygenase as well as peroxidase activity. Consequently, although glutathione peroxidase is capable of reducing 15-hydroperoxy into 15-hydroxy prostaglandin, the participation of this enzyme in prostaglandin biosynthesis is not required.

References

1. Cagen, L.M., Pisano, J.J., Ketley, J.N., Habig, W.H., Jakoby, W.B.: The conjugation of prostaglandin A_1 and glutathione catalyzed by homogeneous glutathione S-transferases from human and rat liver. Biochim. Biophys. Acta 398, 205-208 (1975)
2. Cagen, L.M., Fales, H.M., Pisano, J.J.: Formation of glutathione conjugates of prostaglandin A_1 in human red blood cells. J. Biol. Chem. 251, 6550-6554 (1976)
3. Christ-Hazelhof, E., Nugteren, D.H., Van Dorp, D.A.: Conversions of prostaglandin endoperoxides by glutathione S-transferases and serum albumin. Biochim. Biophys. Acta 450, 450-461 (1976)
4. Christ-Hazelhof, E., Nugteren, D.H.: Purification and characterisation of prostaglandin endoperoxide D-isomerase, a cytoplasmic glutathione requiring enzyme. Biochim. Biophys. Acta (to be published)
5. Flohé, L., Eisele, B., Wendel, A.: Glutathion-Peroxidase, I. Reindarstellung und Molekulargewichtsbestimmungen. Z. Physiol. Chem. 352, 151-158 (1971)
6. Habig, W.H., Pabst, M.J., Jakoby, W.B.: Glutathione S-transferases. The first enzymatic step in mercapturic acid formation. J. Biol. Chem. 249, 7130-7139 (1974)
7. Hamberg, M., Samuelsson, B.: Detection and isolation of an endoperoxide intermediate in prostaglandin biosynthesis. Proc. Natl. Acad. Sci. U.S.A. 70, 899-903 (1973)
8. Hamberg, M., Svensson, J., Samuelsson, B.: Thromboxanes: A new group of biologically active compounds derived from prostaglandin endoperoxides. Proc. Natl. Acad. Sci. U.S.A. 72, 2994-2998 (1975)
9. Johnson, R.A., Morton, D.R., Kinner, J.H., Gorman, R.R., McGuire, J.C., Sun, F.F., Whittaker, N., Bunting, S., Salmon, J., Moncada, S., Vane, J.R.: The chemical structure of prostaglandin X (prostacyclin). Prostaglandins 12, 915-928 (1976)
10. Miyamoto, T., Ogino, N., Yamamoto, S., Hayaishi, O.: Purification of prostaglandin endoperoxide synthetase from bovine vesicular microsomes. J. Biol. Chem. 251, 2629-2636 (1976)
11. Nugteren, D.H., Hazelhof, E.: Isolation and properties of intermediates in prostaglandin biosynthesis. Biochim. Biophys. Acta 326, 448-461 (1973)
12. Ogino, N., Miyamoto, T., Yamamoto, S., Hayaishi, O.: Prostaglandin endoperoxide E-isomerase from bovine vesicular gland microsomes, a glutathione-requiring enzyme. J. Biol. Chem. 252, 890-895 (1977)
13. Simons, P.C., Van der Jagt, D.L.: Purification of glutathione S-transferases from human liver by glutathione-affinity chromatography. Anal. Biochem. 82, 334-341 (1977)
14. Van der Ouderaa, F.J., Buytenhek, M., Nugteren, D.H., Van Dorp, D.A.: Purification and characterization of prostaglandin endoperoxide synthetase from sheep vesicular glands. Biochem. Biophys. Acta 487, 315-331 (1977)
15. Van Dorp, D.A.: Essential fatty acid metabolism. Proc. Nutr. Soc. 34, 279-286 (1975)

Subject Index

Acetamide (halide), glutathione depletion 189
Acetaminophen (paracetamol) 183,189, 194
N-Acetyl cysteine derivatives 49, 158,164,194
N-Acetylation 195
Acetylcarnitine, mitochondria 142
Active site
  γ-glutamyltransferase 46,78
  glutathione synthetase 49
  glutathione peroxidase 107
  glutathione reductase 149
Affinity chromatography 204
Alanylglycine 196
Amino acids
  analysis 60,107,198
  distribution ratio in kidney 84
  glutathione peroxidase 111
  transport 43,60
  uptake by hepatocytes 15
  uptake by renal slice 84
  uptake from tubular lumen 64
p-Aminohippuric acid uptake 85
2-Amino-4-oxo-5-chloropentanoate 48
Aminopyrine 121,164
Anoxia 27,95
Antibodies against γ-glutamyltransferase 67
Antimycin 104,127,196
Arachidonic acid 201
Arrhenius plot, γGT 76
Aspartate and amino acid transport 43,64
ATPase activity 10,84,142
Autoradiography, whole body 37
Autoxidation cysteine 11
Azaserine 46

Benzpyrene and metabolites 160
Benzylammonium-chloride 121,157
Bile
  excretion of glutathione conjugates 120
  release of glutathione disulfide 120
Bilirubin and glutathione transferase 157
Borate 47,79
Bromobenzene metabolism 158,164
Bromosulphthalein 65,140
Bromotrichloromethane 176
Brushborder membrane 54,65,75,87
t-Butylhydroperoxide and GSSG release 97,123
  redox changes 100
  substrate of GSH peroxidase 97,109

Calcium movement in mitochondria 139
Carbon tetrachloride 176
S-Carboxymethyl derivatives
  of cysteine 81
  of glutathione 14
  of selenocysteine 107
Carcinogenesis and glutathione transferase 162
Catalase
  and GSH peroxidase 96
  H_2O_2 elimination 96,166
  intracellular location 96
Cell damage
  and cysteine 10,192
  in glutathione depletion 189
  in tocopherol deficiency 189
Chemiluminescence and lipid peroxidation 102
Chloroacetamide 191
Chloroacetate 107
1-Chloro-2,4-dinitrobenzene 116,120, 158,205
Collagenase 8,14
Cumene hydroperoxide
  glutathione release 97
  substrate of GSH peroxidase 97,114
Cyclooxygenase, prostaglandin 201
Cystathionine, pathway 8,13,172
Cysteamine 150
Cysteine 9,150
  cell damage 10,189,192
  content of liver 3,25
  and glutathione content 8,13,171
  and glutathione peroxidase 112
  and glutathione transferase 159
  half-life of, 6
  metabolic fate 4
  reservoir of, 3
  sulfur-sulfur-bound 148
  uptake 15
Cysteine-glutathione disulfide 15,150, 198
Cysteine homocysteine disulfide 18
Cysteinylglycinase 43,194
Cysteinylglycine 43,196

Cystine
  dimethylester 160
  uptake 15
Cytochrome C peroxidase 100
Cytochrome P_{450} and glutathione 11,164

Deficiency
  γ-glutamylcysteine synthetase 50
  γ-glutamyltransferase 50,66
  glutathione 189
  glutathione synthetase 50
  selenium 114,120
  tocopherol 95
Demethylation of drugs 122,164
Detoxification
  bromobenzene 164
  diabetes chronic 22
  epoxides 164
  formaldehyde 164
  glutathione transferase 157,164,189
  hydroperoxide 164
  lipid peroxides 176,183,189
Diamide
  amino acid uptake 84
  ATPase activity on mitochondria 142
  calcium ion movement in mitochondria 139
  gluconeogenesis 84
  glutathione content in mitochondria 127,140
  and GSH-content in kidney 84
  magnesium ion movement in mitochondria 139
  sugar uptake 84
  and thioloxidation 140
Diazenes, reduction in mitochondria 127
6-Diazo-5-oxo-L-norleucine 22,46,78
Dichlorophenol indophenol 133
Dichloronitrobenzene 160
Diethylmaleate 170,186,190
Difference spectroscopy of glutathione peroxidase 108
L-dihydroorotic acid 48
2,3-Dimercaptopropanol 87
DIP 128
DIP^{2+} 128
Disulfides 151
  metabolism 27,127,148,194
Dithioerythritol 34,62,151
Dithiothreitol 9,85,143,151
  influence on glutathione assay 33
  influence on glutathione efflux 34
Drug oxidation 120,164,194
Drugs and lipid peroxidation 164,183

EGTA 8,139
Electrophilic conjugation 157,164,195
Endoperoxides 201
Epoxides 148,164
Equilibrium
  5-oxoprolinase 46
  thiol-disulfide interchange 20,148
Ethacrynic acid 160
Ethane formation 95,176,183
Ethanol and lipid peroxidation 176
Ethionine 9
N-Ethylmaleimide 127
Ethylmorphine 122,164

Ferritin 67
Fetal mouse liver 22
Fluorodinitrobenzene 14
Formaldehyde dehydrogenase 164
Formaldehyde and glutathione 164
Formylglutathione 168
  and ethane formation 177

Gas chromatography 176,183
Glomerular filtration 29,52,84
Gluconeogenesis, renal slice 84
Glucose-6-phosphatase 87
Glutamate 8,43
  reabsorption in kidney 43,63
  in liver and kidney 11
Glutaminase 61,70
Glutamine 8,45
γ-Glutamyl cycle 39,43,60,70,194
  amino acid transport 43,63
  glutathione metabolism 32,43
γ-Glutamylcyclotransferase 43
  substrate specificity 48
γ-Glutamylcysteine synthetase 11,43
  deficiency 50
  inhibition 48
  substrates 48
γ-Glutamylpeptides 43,60,78,194
γ-Glutamyltransferase
  active site properties 46,78
  activity, effect of temperature 70
  antibodies 60
  chronic diabetes, effect on 22
  deficiency 50,66
  degradation of glutathione conjugates 60,194
  degradation of glutathione (glutathionase) 39,60,70,194
  distribution 44,60
  function 44,60,70,194
  γ-glutamyl-enzyme intermediate 46,78
  increase of activity 22
  inhibition 22,46,78,87
  interaction with membranes 70
  papain-purified 46,70
  pregnancy, effect on 22
  reaction, types of 46,60
  starvation, effect on 22
  topology in kidney 60
  Triton-purified 46,70

γ-Glutamyltranspeptidase, see
γ-Glutamyltransferase
Glutathione
and amino acids 50
arterial plasma 28
assay dithiothreitol, influence on 33
biliary 120
binding to glutathione peroxidase 108
binding to glutathione transferase 157
biosynthesis 18,172,190
catalytic role 148,158,168,201
in caval perfusate 122
content in liver 3,9,18,151,167
cysteine, effect on 8,171
food intake 3,193
hepatocyte preparation, effect on 8
homocysteine, effect on 9
methionine, effect on 9,13,171
pregnancy 23
starvation 22,183
tryptophan, effect of 4
content in mitochondria 127,140
diazenes, effect on 127,140
hydroperoxides, effect on 127
inhibitors, effect on 127
uncouplers, effect on 127
and cytochrome P-450 164
depletion of 164,183,189
depletion and cellular function 189
detoxification of
bromobenzene 164
epoxides 164
formaldehyde 164
hydrogen peroxide 164
lipidperoxides 175,183
and drug oxidation 120,164,183,194
extracellular 17,32,60,167
extracellular hydrolysis 32,60,70,78
half-life in nephron 60
and hydroperoxides 97
increase 4
metabolism in kidney and liver 32,43, 60,164,194
oxidation 100,139,194
permeability 34
and phenobarbital treatment 20,164,194
pool "labile" 3,6,27,187
and prostaglandin metabolism 201
and protein kinase activity 86
and pyridine nucleotide oxidation 100,167
regeneration in mitochondria 127
release 18,27,32,34,95,118,120
removal from plasma 32
renal venous plasma 28
and selenium deficiency 120
status of 17
and sulfur-sulfur bonds 148
transport of 27,32
transport activity 84
turnover 27,183,189,194
uptake by kidney 27,60
Glutathione S-conjugate
metabolism 194
and γ-glutamyl-transferase 194
release 120
Glutathione disulfide (GSSG)
in bile 120
metabolism in kidney and liver cells 15,194
release from perfused liver 126
Glutathione oxidising agents
cation gradients 87
Na^+ ,K^+-ATPase 87
Glutathione paracetamol conjugate 194
Glutathione peroxidase
localisation 96,127
peroxide elimination 96,164,176,183
Glutathione peroxidase, Selenium dependent
active site properties 107
difference spectroscopy 108
distribution 116
Selenium moiety 107
substrate binding 107
Glutathione peroxidase, Selenium independent 114
distribution in organs 114
distribution in species 114
function 118
glutathione release 118,164
and glutathione S-transferase 114, 159,194
lipid peroxidation 176,183
and peroxide metabolism 114
properties 115
and prostaglandin endoperoxide metabolism 201
purification 115
substrates 114
Glutathione reductase
glutathione reduction 100
in mitochondria 127
pyridine nucleotide oxidation 100
scission of sulfur-sulfur bound 148
specificity 149
Glutathione synthesis 4,8,13,23,32,43
Glutathione synthetase(s) 23,43
deficiency 50
substrate specificity 48
Glutathione-S-transferase (=ligandin) 120
bilirubin 157
carcinogenesis 162
detoxification 157
distribution 157
glutathione binding 157
ligandin 161
ligands 161
and Non-Selenium dependent glutathione peroxidase 114,159
products 158

Glutathione-S-transferase (=ligandin)
  and prostaglandin endoperoxide
   metabolism 201
  purification 158
  reaction mechanism 157,158,160
  reaction, types of 157
  scavenger role 161
  sheep lung 204
  storage protein 157
  substrates 160
Glycine 8,13,32,43,60,84
  permeability liver 35
  uptake kidney 60
Glycocalix 65
Glycolate 99
GSH, see glutathione
GSSG, see glutathione disulfide

Half-life
  glutathione liver 3,27
  kidney 39,43,60
Haloacetates
  glutathione depletion 189
  glutathione peroxidase 110
  γ-glutamyltransferase 79,80
Halogenated hydrocarbons 176
Halothane 176
Hepatocytes 8,13,32,164,189,194
High-energy intermediates,
   mitochondria 134
Homocysteine 9,150
HPLC analysis of sulfur aminoacids 13
Hydrogen peroxide
  elimination
    by catalase 96,166
    by GSH peroxidase 96
  formation
    in hepatocytes 8,99,164
    in perfused liver 99
  and microsomes 164
  oxidation of GSH 164
  substrate of GSH peroxidase 97,
   114,164
Hydroperoxide and glutathione release 95
Hydroperoxide metabolism 20,95,114,164
Hydroperoxides
  and pyridine-nucleotides 100
  reduction in mitochondria 127
  as substrates for GSH peroxidase
   97,114
15-Hydroperoxy-prostaglandin 201
p-Hydroxymercuribenzoate 127,204
S-Hydroxymethyl-glutathione 168
Hyperbaric oxygen 95

L-2-Imidazolidone-4-carboxylate,
   inhibition of 5-oxoprolinase 48
Immunochemical localisations γ-GT 60
Inhibitors
  γ-GT 22,32,47,79
  5-oxoprolinase 48
  γ-glutamylcysteine synthetase 48
  glutathione synthetase 49
Insulin, scission of sulfur-sulfur-
   bounds 150
Inulin clearance kidney 60
Iodoacetamide
  γ-glutamylcyclotransferase 47
  glutathione depletion 190
  glutathione peroxidase 110
Iodomethane 160

Kidney
  glutamate reabsorption 63
  γ-glutamyl-cycle 43
  γ-glutamyltransferase, localisation
   60,70
  glutathione degradation 32,60,70
  glutathione metabolism 3,8,13,27,32,
   43,60,120,157,164,189
  glutathione removal from plasma 32,37
  glutathione uptake 27,43,84
  non-Se-glutathione peroxidase 117
  peptide bond synthesis 11
  role of glutathione in transport 60,84
  tubules 38,60
Kinetics
  γ-GT 46,70,79
  S-transferases 158
  thiol-transferases 150
Krebs cycle intermediates and gluta-
   thione 129

Lactate 8
Lecithin vesicles 71
Ligandin, see glutathione-S-transferase
Linoleic acid
  and lipid peroxidation 180
  and prostaglandins 201
Lipid peroxidation
  carbon tetrachloride 176
  cell damage 189
  chemiluminescence 102
  and drugs 183
  ethane exhalation 176,183
  ethanol 176
  glutathione peroxidase 183
  and glutathione peroxidase activ-
   ity 176
  halogenated hydrocarbons 176
  of membranes 176
  in microsomes 176
  oxygen concentration 176
  starvation 183
Lipid peroxide formation 95,176,183,189
Lipoamide dehydrogenase, specificity 149
Liver necrosis and paracetamol 183
Lysine 9

Magnesium movement in mitochondria 139
Malonate 127
Malondialdehyde
  formation from peroxidized lipids 179,183,189
  glutathione depletion 189
Membrane
  association of γ-glutamyltransferase 60,70
  lipids 176
  mitochondrial 139
Mercaptoethanol 85
α-Mercaptopropionylglycine 189
Mercapturic acids 43,60,157,164,194
Metalloflavoproteins and superoxide anions 98
Methionine 9,16,47,85,172
  glutathione content 171
  utilisation in glutathione depletion 189
L-Methionine sulfoximine phosphate 48
S-Methylcysteine 64
α-Methyl-D-glucoside 85
S-Methylglutathione 64,204
Methyliodide 62
Microperfusion renal tubule 60
Microsomes
  and catalase 95
  and γ-GT 43,70
  lipid peroxide formation 176
  mixed function oxidase 164
  protein bound metabolites 173
Mitochondria, rat liver
  chemiluminescence 104
  content of glutathione 127,140
  diazene reduction 127
  functions 139
  glutathione oxidation 139
  glutathione peroxidase 127
  glutathione reductase 127
  high energy intermediates 134
  hydroperoxide reduction 127
  permeability 136,139
  regeneration of glutathione 127
  swelling 143
  thiol groups 139
Mixed disulfides
  formation in mitochondria 142
  metabolism 198
  with proteins 148
Mouse fetal liver 22
Mouse liver 183

NADPH, oxidation
  and glutathione release 100,164
  and hydroperoxides 100
  in mitochondria 127
Na^{+},-K^{+}-ATPase 87
Nephron, topology 60
N-ethylmaleimide 133
p-Nitrophenylacetate 160

Oleate 8
Oligomycin 134
Ophthalmic acid 37,60
  half-life in nephron 60
Organic peroxide formation 95
5-Oxoprolinase 49
  inhibition 48
  substrate specificity 48
Oxygen
  hyperbaric 95
  intermediates 95
  and lipid peroxidation 176
  tension and hydroperoxide formation 95
  toxicity 95

Papain-purified γ-glutamyl-transferase 70
Paracetamol (acetaminophen) and glutathione 183,189,194
Paracetamol-N-acetylcysteine 194
Pentane, exhalation 180,184
Peptide bond synthesis 10
  methionine, effect on 11
  serine, effect on 11
Perfluorohexane 167
Permeability of glutathione 34,60
Permeability of oxidants into mitochondria 136,139
Peroxide formation 95
  metabolism 114
Phenobarbital 120,167,194
Phenylarsenoxide 133
O-Phthalaldehyde method 196
Platelet aggregation 201
Polyacrylamide gel electrophoresis
  γ-GT 72
  PG-isomerases 205
Pregnancy and γ-GT 22
Prostacyclin formation and glutathione 202
Prostaglandin E-formation 201
Prostaglandin endoperoxide D-isomerase from rat spleen 201
  inhibition 204
  properties 204
  purification 203
  specificity 204
Prostaglandin endoperoxide E-isomerase 201
Prostaglandin endoperoxide metabolism 201
  and glutathione requiring enzymes 201
Protein kinase activity 86
Protein mixed disulfides 148
Proteins
  dietary influence on glutathione 3
  disulfide-containing 151
  SH-storage protein 157
Proteolysis, effect on disulfide reduction 152

Prothionine sulfoximine 53
Proximal convolution 43,60,70
Pyridine nucleotide oxidation
  and glutathione release 100,164
  and hydroperoxides 100

Release glutathione liver 27,34
  hepatocytes 8,189,194
Renal plasma flow 29
Ribonuclease 150
Rotenone 134,144
Ruthenium red 139

Scavenger role of glutathione
  transferase 161
Selenium
  content of GSH peroxidase 107
  deficiency 114,120
  -independent glutathione
  peroxidase 114
Serine 10,47,82
Singlet oxygen 97
Starvation 22,183
Storage, protein(s), glutathione
  transferase 157
S-Substituted glutathione derivatives
  43,60,120,157,194
Substrate analogs γ-glutamylcycle 49
Succinate
  and mitochondrial glutathione 104,
  127,143
Sucrose
  loaded vesicles 75
  uptake by perfused liver 32
Sugar uptake by renal slice 84
S-Sulfocysteine 150
S-Sulfoglutathione 150
Sulfur-sulfur bonds
  and glutathione 148
  glutathione reductase 148
  lipoamide dehydrogenase 149
  scission of, 148
  thioltransferase 148
  thioredoxin 149
Superoxide dismutase 95,168

Thenoyltrifluoroacetone 134
Thiobarbituric acid 28,164,183,189
Thioethers 158
Thiol agents, effects on mitochondrial
  glutathione 132
Thiol-disulfide interchange 18,148,194
Thiol groups
  content in mitochondria 140
  in membranes 139
  oxidation in mitochondria 139
  redox state in mitochondria 139
  and sulfur-sulfur bonds 148
Thiol transferases
  and glutathione reductase 150
  insulin 150
  kinetics 150
  localisation 148
  properties 150
  scission of sulfur-sulfur bonds 148
  substrates 150
Thioredoxin, specifity 149
Thromboxanes, A_2 and B_2 formation and
  glutathione 202
Tocopherol deficiency 95
Tocopherol and hydroperoxides 98,189
Transhydrogenase 137
Transport
  GSH 32,84
  amino acids 43,60,84
Transsulfuration pathway 191
Trichloroethylene 176
Trinitroglycerol 160
Triton X-100 and γ-GT 70
Tryptophan 4

Uncouplers and mitochondrial gluta-
  thione 104,127
Uptake
  amino acids kidney 60
  glutathione kidney 84
  glutathione liver 32

Vinylchloride 176

MIX
Papier aus verantwortungsvollen Quellen
Paper from responsible sources
FSC® C105338
FSC
www.fsc.org

If you have any concerns about our products,
you can contact us on
ProductSafety@springernature.com

In case Publisher is established outside the EU,
the EU authorized representative is:
Springer Nature Customer Service Center GmbH
Europaplatz 3, 69115 Heidelberg, Germany

Printed by Libri Plureos GmbH
in Hamburg, Germany